MW01670716

PCI™ Precast/Prestressed Concrete Institute

Erector's Manual

Standards and Guidelines for the Erection of Precast Concrete Products

MNL-127-99

ERECTORS' MANUAL

STANDARDS AND GUIDELINES FOR THE ERECTION OF PRECAST CONCRETE PRODUCTS

prepared by

PCI ERECTORS COMMITTEE
MICKEY PICKELL, Chairperson

MARK BIEBIGHAUSER
DAVID BUESING
SAM CITRO
STEVE COOPER
DONALD DICKIE
JOSEPH DUGAN
WAYNE EDWARDS
ALVIN ERICSON
SKIP FRANCIES
SID FREEDMAN
GREGORY GIBBONS
GERALD W. KRIEGEL
JARED LANDWEHR
CHARLIE LeMASTER
CHARLES N. MAYER
WILLIAM R. MILLER
JAMES MIRGLIOTTA
RAYMOND ROBERTSON
DANIEL R. ROTEN
JOHN SCHNELL
JIM SHULTZ
ASHLEY TOWNSEND
ROBERT WEISS
KENNETH WENRICH
RICHARD K. WESEN
MIKE WHITE
GARY N. WILDUNG
S. ALAN WILLIAMS

MNL 127-99

Copyright © 1999
By Precast/Prestressed Concrete Institute

All rights reserved.
This book or any part thereof may not be reproduced
in any form without the written permission of the
Precast/Prestressed Concrete Institute.

Substantial effort has been made to ensure that all data and information in this manual are accurate. However, PCI cannot accept responsibility for any errors or oversights in the use of material or in the preparation of engineering plans. This publication is intended for use by personnel competent to evaluate the significance and limitations of its contents and able to accept responsibility for the application of the material it contains. Special conditions on a project may require more specific evaluation and practical engineering judgment.

While every effort has been made to prepare this publication as the national standard for the industry, it is possible that there may be some conflicts between the material herein and local practices.

First Edition, 1985
Second Edition, 1999

ISBN 0-937040-56-8

Printed in U.S.A.

ERECTORS' MANUAL

Standards and Guidelines for the Erection of Precast Concrete Products

TABLE OF CONTENTS

INTRODUCTION

This manual is intended to provide **Standards** and corresponding guidelines for the erection of plant-produced precast and prestressed concrete products, and to establish a level of quality that will be recognized and respected by the construction industry.

The development and formatting of this manual has been guided by the PCI Plant Certification Program to allow this document and the PCI manual, "Erection Safety for Precast and Prestressed Concrete" (MNL-132) to be the primary source documents for the implementation of a quality assurance program for erectors. This manual is intended to describe a set of satisfactory quality control procedures for a general precast concrete erection operation. It should be augmented as required for specific operations and products by the specifier or erector.

This manual includes **Standards** shown in boldface type. These are the important fundamental requirements governing the quality of installed precast and prestressed concrete elements. The remainder of the manual is the Commentary which amplifies and explains the **Standards.** The Commentary also provides many suggestions regarding how to perform various erection operations.

Tolerances provided are industry tolerances developed by the PCI Tolerance Committee. These tolerances were developed with extensive input from designers, precasters and erectors.

A successful precast concrete project requires teamwork. This means close cooperation and coordination of all participants including the owner, architect, engineer, precaster, erector, general contractor and/or construction manager (GC/CM), and all other affected trades. The basic working relationship between these parties must be established at an early stage in the development of a project in order to achieve the desired schedule and results. Each of these parties has the responsibility for communicating with the other parties through the GC/CM in order to achieve the best efficiency during construction and to provide a timely completed quality structure.

Quality procedures during erection are essential. The work of the precast concrete erector results in one of the most visible elements of a construction project. Even though the quality of the precast concrete components delivered to the jobsite may be outstanding, the success of the entire project is finally dependent upon the quality of workmanship by the erector. A poorly erected project can detract not only from the performance of the structure, but also from the appearance of even the best designed precast concrete structure. The erector should assist the GC/CM in taking the lead on a construction site to help solve problems before they occur through the coordination of work of all relevant participants. Foresight and planning will help keep the project moving on a trouble-free course.

It is also very important that the erector, whether working directly for the precaster or not, recognizes that they are an "ambassador" for the precaster as well as the precast industry as a whole. The erector has more inter-relationship with the "customer" than anyone else representing the precaster, and should therefore deal with the construction team in a tactful and professional manner to help bring the project to a successful conclusion.

The contractual responsibility for the erection of precast concrete may be established through one of the following methods:

1. The precaster supplies the product erected, either by in-house labor or through an independent erector.
2. Erection is accomplished by either the GC/CM or by an independent erector under a separate agreement with the GC/CM and/or owner. The precaster is responsible only for supplying the product, F.O.B., plant or jobsite.
3. The products are purchased by an independent erector who enters into a contract with the GC/CM and/or owner to furnish and erect the complete precast concrete structure.

On projects where the precaster does not have the contractual obligation for the erection, it is very important that they assign a representative to observe, coordinate and report on erection activity. Regardless of contractual obligations, it is recommended that the precaster maintain sufficient contact with the firm(s) responsible for both transportation and erection to ensure that the precast concrete units are timely delivered, properly handled and erected according to the design and project specifications and this manual. Erection of architectural and structural precast concrete should be performed only by qualified erectors employing workers who are properly trained to safely handle and install the product.

DIVISION 1 — PRECONSTRUCTION PLANNING

1.1 Purpose

The probability for success in a construction project begins well before any work has started at the site. One of the most important steps in a project is preconstruction planning. It is an activity that is very often overlooked. It is at this stage that the "seeds" are planted for success or failure.

Preconstruction planning implies "building the job in your mind" and anticipating as many factors as possible that could affect the work. There is no substitute for experience in predicting jobsite difficulties or constraints; however, the purpose of this section is to guide the precaster and/or erector in areas that should be considered and addressed. The areas of concern that should be addressed during preconstruction planning are:

1. Determining site access, erection direction and sequencing.
2. Identifying hazards.
3. Determining size and weight limitations.
4. Erection planning.
5. Equipment selection.
6. Developing an erection safety plan.
7. Field layout and verification.
8. Special handling requirements.
9. Special lifting hardware.

This order indicates the usual sequence of activities but many factors occur that influence project planning. Each step is further considered in this chapter.

1.1.1 Responsibility of the General Contractor

The general contractor or construction manager (GC/CM) is normally responsible for project schedule, grid dimensions, and coordination with all construction trades. Relative to erection of precast concrete, the GC/CM in conjunction with the precaster should:

1. Be responsible for coordinating information necessary to produce the precast concrete erection drawings. A preconstruction conference should be held as soon as possible after award of the precast concrete contract. This meeting should consider all details covered in this section, including jobsite access, erection sequencing, weight/size limitations, special rigging and guy/bracing scheme information. The sequencing and guying/bracing information should be communicated to the precast engineer. If necessary, the precast engineer would then develop, in conjunction with the erector and engineer of record, a bracing sequence to maintain stability of the structure during erection. Limitations may state, for example, that loading of the structure be balanced, requiring that no elevation be erected more than a stated number of levels ahead of the remaining elevations. Other limitations may involve the rigidity of the structure, requiring that walls not be erected prior to completion of floors designed to carry the lateral loads.

   For steel frames, it should be determined how far in advance the final connections of the frame are to be completed prior to panel erection. Particular consideration should be given to deflection and/or rotation of the supporting steel members due to the precast concrete and other superimposed loads. For concrete frames, the GC/GM should furnish to the erector, strength of concrete and authorization to begin erection.

   The engineer of record should also recognize that connections between panel and frame impose concentrated loads on the frame and that these loads may require supplemental reinforcement. In the case of multi-story concrete frames, consideration should be given to the effects of frame shortening due to shrink age and creep.

2. Review and approve (or obtain approval) for all erection drawings and related design.
3. Be responsible for the coordination of dimensional interface between precast concrete and the work of other trades.
4. See that proper tolerances are maintained by other trades to guarantee accurate fit and overall conformity with precast erection drawings.

5. Be responsible to provide and maintain clear, level, well-drained unloading areas and road access around and into the structure sufficient so that the hauling and erection equipment for the precast concrete units are able to operate under their own power. Units should be located to permit lifting by the erection equipment from the trailers and placement directly into the structure.

1.2 Determining Erection Direction and Sequencing

1.2.1 General

This part of preplanning depends upon the type of structure being erected (e.g., one level of hollow-core deck, a multi-level parking structure, architectural cladding, etc.). Site access and conditions will also be a factor in developing the direction of erection. Proper planning of the construction process is essential for efficient and safe erection. **The sequence of erection shall be established early and accounted for in the bracing analysis and erection drawings.**

On multi-story cladding projects, the erector should determine how many times the erection equipment must be moved to complete the work. Also, on steel frames, the timing of application of fireproofing should be considered for prewelding that must occur. On concrete frames, the location of formwork and any overhanging scaffolding will sometimes influence the method of erection and sequencing.

1.2.2 Direction of the Work

The erector shall determine, together with the GC/CM and precaster, the conditions that will influence the direction of the work. Often, certain areas of the building will take precedence over others due to extensive utilities and/or finish work. At other times, the location of existing buildings, temporary works, landscaping, terrain features or building stability will dictate the flow of work. The construction schedule should be planned so that chases, conduits, pipes, drains, or masonry work do not interfere with the erection of the precast concrete units. Otherwise damage to these items may result when the precast concrete units are placed.

1.2.2.1 Site Access. Consideration shall be given to access for erection and delivery equipment. At times, the direction of erection will be determined by gate entry points into the jobsite or the location of construction trailers and storage facilities. The GC/CM should indicate where materials will be stored and man-hoists will be located. Other jobsite factors, including disassembly of the crane, will affect access and the direction that work must proceed. Free and easy access to appropriate parts of the structure can many times present the single largest challenge to the erector. **Site access shall be a provision of the contract.**

1.2.3 Determining and Following Sequence

The sequence is probably the most important factor influencing the erection of the structure. **Prior to the manufacture of precast concrete members, the erection and job sequence shall be agreed upon by the GC/CM, precaster and erector.** As stated earlier, every effort should be made to allow a unit to be handled in one motion from unloading to final position. Multiple handling results in excessive cost and possible delays. **When a sequence is agreed upon, it shall be strictly followed until conditions require a mutually agreeable change.**

The sequence of erection is controlled by many factors. The most significant ones are crane accessibility, building structural form (plane shape, unbraced/braced frame) and locations of walls for stability.

Sequence for the erection shall be considered when selecting the crane size. Proper sequencing minimizes the number of obstructions and thus enables selection of a crane with the shortest possible boom length. Economy is achieved by placing (spotting) the crane at locations from which the greatest number of units can be lifted and placed before further crane moves are needed. In parking structures, another common strategy is to start at one end of the building, build to full height, and move the crane out as the structure is completed.

Many projects require only one move-in for erection. Others are erected in two or more phases, possibly days, weeks or months apart. With proper and timely information, the erector can more easily schedule crews and equipment for the project. This makes it important that the erector be provided accurate schedule information early in the project. The information must be updated whenever it changes.

A careful review of all of these factors will allow the precaster/erector to work in harmony with the construction site constraints and the needs of the owner and GC/CM.

1.3 Identifying Hazards

A careful review and inspection of the jobsite shall be performed to identify obstructions that could limit equipment mobility.

Overhead obstructions shall be carefully considered in planning. Power lines may have to be moved or deactivated. Removal of tree limbs may be necessary.

All existing property to be protected, such as underground utilities, tanks, tunnels, trees shrubs, existing asphalt and concrete shall be clearly identified.

In projects that are in close proximity to airports, lights and/or flags may be required on booms.

1.4 Precast Size and Weight Considerations

Product size and weight constraints can be affected by a number of factors. These include available erection equipment, jobsite conditions or the precaster's production equipment or transportation limitations. It is of the utmost importance that the erector's weight limitations be considered by the precaster and precast engineer as the erection drawings and product designs are being developed.

One limitation is whether the erector has the capability of rotating pieces (i.e., does the crane have one line or two). This could dictate how the pieces are shipped to the jobsite and whether additional or special handling devices are required.

Another factor is panel shape. Odd-shaped panels may require special erection considerations and therefore require the erector to provide the precast engineer with information about special handling of odd-shaped pieces.

To avoid unrealistic or impossible erection conditions, a review of the project by the architect/engineer with the precast engineer and erector is beneficial. **The precast engineer and erector shall determine the requirements for handling and erecting the members.** Safety and efficiency should always be considered. The architect/engineer should give careful consideration to all of the factors that affect the methods that can be used in the construction of the project and prepare the design to best suit jobsite conditions.

1.5 Erection Planning

The erector should maintain a checklist of items that have been shown to help expedite projects. Some of these items concern site conditions and relationships with the GC/CM and other trades. Others deal with the design of the product and connections. Erectors should consider at least the following items:

1. Verify the availability of products.
2. Check that erection drawings are easy to understand, properly cross-referenced, and accurate.
3. Suggest connections that are safe and allow quick release of the crane hook. This would require mechanical, bolted connections which allow adjustment of the position of the precast piece after the crane is released and sometimes additional temporary connections.
4. Check for grouted connections in cold climates which require special planning and procedures.
5. Determine if loose hardware, erection materials and bearing pads are available to be delivered to jobsite.
6. Prevent confusion by minimizing the number of different types of erection hardware. Make the designer aware of this important issue, particularly for those materials that vary only slightly.
7. Compile area by area hardware breakdowns and summaries.
8. Verify that attachments are accurately placed in the structure.
9. Verify that the precast concrete is dimensionally correct (i.e. within tolerance). Plates, sleeves, inserts and openings should be checked and cleaned.
10. Review that connections are designed to industry standard tolerances and are placed at convenient locations.
11. Plan the fewest number of pick points with one type of lifting device or as few different types as possible to avoid rigging changes.
12. Whenever possible, have units shipped in the orientation in which they will be installed (i.e., walls vertical, floor slabs horizontal, proper position end-for-end). Some long members must be oriented properly on the truck because of site or rigging restrictions. If vertical pieces are too large and must be shipped flat, ensure that panels are designed so they can be rotated.
13. See that consideration for fall protection is designed into the structure, with tie-off eyes or inserts where needed. Holes in floor and roof members and openings in walls should be covered while in the precast concrete plant.

1.6 Equipment Selection

There is no single crane suited to all precast concrete field operations. Size and weight of product, location of precast piece within the structure, size and shape of structure, access, overhead obstructions, and individual crane capacities all contribute to unique requirements for the crane. **Selection of the optimum crane shall be decided by a systematic process involving the following steps:**

1. As early as possible, the erector should obtain a site plan showing existing jobsite conditions as well as grades that will exist at the time of installation. By reviewing this plan, the erector can identify in advance conditions which may present access or crane reach problems. **Prior to erection, a site inspection and meeting shall be held to confirm that the necessary access routes and crane set-up locations are available.** Based on these locations the maximum working radius is determined. This working radius, or reach, is measured from the crane center pin to the center of gravity of the farthest pick.

2. **The maximum hoisting weight shall be determined by adding the piece weight to the weight of all rigging hanging from the boom tip. This rigging shall include the weight of jibs, hooks, hook blocks, slings, and all other load-handling devices including the hoist rope.**

3. **Obstructions both above and below the boom travel area shall also be noted.**

4. **Physical data including maximum reach and available boom travel area shall be plotted on the range diagram for the proposed crane to determine working boom angles.**

5. **Maximum pick and boom angle limitations shall be traced on the rated lifting capacity chart to determine the safe lift capacity.** If this rated lift capacity exceeds the computed hoist weight, the crane investigated can safely erect the job. If the rated lift capacity is less than the computed hoist weight, a larger crane will be required.

Additional information concerning equipment selection can be found in Chapter 3.

1.7 Developing an Erection Safety Plan

A specific jobsite Erection Safety Plan shall be developed. An Erection Safety Plan includes some or all of the following:

1. Erection drawings — These are part of the Plan when they address the stability of the building or components, and/or when they include temporary connections, and shoring, guying, bracing, and sequence requirements during the erection process.

2. Company and/or industry safety manuals — These should include specific provisions for the safety of employees and address precast concrete erection hazards such as those requiring fall protection.

3. A Fall Protection Plan specific to the project.

For further guidelines on an Erection Safety Plan refer to the PCI manual, "Erection Safety for Precast and Prestressed Concrete" (MNL-132).[17]

1.8 Field Verification and Layout

Layout shall be required before and during erection.

1. **Before erection so that the starting point is clearly established and any possible tolerance conflicts with the support structure are determined early.**

2. **During erection to maintain proper alignment.**

The benefits of layout prior to erection are: (1) when an error is found it can be corrected prior to starting erection; (2) costly erection delays for crane and crew are avoided when everything fits; and (3) layout crews may find errors on the shop drawings which can reduce costly corrections and minimize delays.

It is the responsibility of the GC/CM to establish and maintain, at convenient locations on the foundation or the building, control points, benchmarks and lines in an undisturbed condition for use by the erector until final completion and acceptance of the project, Fig. 1.8(a) and (b). If the building frame is not precast concrete, then the benchmarks and building lines should be provided by the GC/CM in every area or at every level. Work points should also be provided for angled and curved building elevations.

The erector should be advised by the GC/CM of any known discrepancies in field location, line or grade, and be provided with suitable work areas for layout.

Prior to start of erection and scheduling of delivery and lifting equipment, a field check of the project should be made by the erector to ensure that foundations, walls and structural frame are suitably constructed to accept the precast concrete units. This should include a field check of the work affecting the erection contract, including location, line and grade of bearing surfaces, notches, blockouts, anchor bolts, field-placed or contractor's hardware

Fig. 1.8(a) Cassion cap showing layout for walls to be erected; (b) Closeup of joint lines.

(miscellaneous iron), a check on dimensional tolerances. The goals of this survey are to ensure that the areas to receive the precast concrete are ready and accessible to the erector and that the precast concrete units will fit. **All dimensions and details taken shall be checked against erection drawings.**

The erector shall confirm that the minimum concrete strength has been achieved.

Notes shall be kept of all discrepancies which exceed specified tolerances. Discrepancies between site conditions and the erection drawings, which may cause problems during erection, such as structural steel out of alignment, anchor bolts or dowels improperly installed, errors in bearing elevation or location, and obstructions caused by other trades, shall be noted in writing and sent, if appropriate, to the precast concrete manufacturer and/or GC/CM. Erection should not proceed until discrepancies are corrected by the GC/CM, or until erection requirements are modified and reviewed by the design engineer. Verification of remedial work should be the responsibility of the GC/CM. **The erector shall return to the site for final layout after all necessary corrections of site conditions in the area where erection will occur have been made by the GC/CM and before any erection is started.**

Working from the control points, benchmarks and lines established by the GC/CM, the erector should establish offset lines, and elevation marks, as required, for use at each floor level, Fig. 1.8(c). It is the layout person's duty to mark the locations to receive the precast concrete units to enable the erector to place them with a minimum of measuring and moving to final position.

Fig. 1.8(c) Layout crew establishing offset lines for architectural cladding on a steel structure.

The erector should establish joint locations prior to actual product installation. This will keep the differential variation in joint width to a minimum, as well as identifying problems caused by building frame, columns or beams that are out of dimensional or alignment tolerance.

The layout crew should be kept well ahead of erection so that if corrections are required, adequate time is available for the design, approval, fabrication, and installation of the repairs. This is necessary so as not to delay the process of erection.

1.9 Types of Layouts

1.9.1 Floor and Roof Systems

The simplest layout is for the placement of hollow-core slabs, double tees, or beams on block walls, grade beams, steel or concrete frame structures. In many cases this layout is accomplished by the foreman and entails only the measuring of side walls and properly locating floor penetrations. In this type of structure no vertical control is used other than to verify bearing elevations and provide a visual check to ensure adequate bearing surfaces.

1.9.2 Single-Story Panels, Columns and Beams

This layout shall include both vertical and horizontal control performed during layout. All precast concrete base locations should be marked. Shims for control of elevation should be set by instrument. The height of shims should be written on the base support and the shim location marked. The horizontal control lines may be under the precast concrete units, along the face, or may be offset. Vertical control lines may be located on the wall or foundation supporting the unit. Miscellaneous erection hardware should be properly located and installed.

Prior to erecting columns, an accurate layout shall establish column locations with respect to building grid lines and base elevations. Elevations may be established at the top of a shim pack or with leveling nuts on anchor bolts.

1.9.3 Multi-Story Total Precast Structures

This type of structure shall receive special care and accuracy during layout. All control lines should be established on the foundation or floors of multi-story buildings in sufficient detail to allow proper installation of the precast concrete units. **The elevator shaft shall be checked on every lift due to the requirement for close tolerances.**

The following procedures are guidelines that should be used for a typical floor:

1. From the primary project benchmark, a benchmark should be established in the elevator shaft or any shaft or column that will be clear and unobstructed for the duration of the erection. From this new benchmark establish the individual floor benchmarks.
2. Establish a building centerline on each floor.
3. Establish offset panel lines. Offset lines for panels should be marked at each end of the wall.
4. Establish shim stack thickness for two points on each wall panel.
5. During wall panel erection once the panel is on the shims and not yet unhooked, check for elevation and plumbness.
6. The joint layout of wall panels, the vertical alignment and a building elevation check should be made at horizontal joints to identify any problems caused by the building frame, columns or beams which are incorrectly dimensioned or out of alignment.
7. Prior to erecting floor members, wall alignment and lateral dimensions should be checked.

1.9.4 Multi-Story Cladding and Bearing Walls

Vertical and horizontal control shall be maintained within tolerances. Elevations should be clearly marked on the supporting structure. On steel or cast-in-place concrete structures, offset lines may be set on the floor slabs or marked on the steel to provide control for the exposed face. Panel joint centerlines should also be located. As erection proceeds, a check should be made on the overall elevation and plumbness of the precast concrete units already placed. However, because of allowable tolerances for steel or concrete frames, it may be necessary to permit cladding to be installed slightly out of true vertical plane. A straight line should be maintained along the outside face of the wall panel. **The critical elevation (primary control surface), as indicated on the erection drawings shall be maintained.** If more than one critical line is indicated, the erector should balance any deviations between the two edges.

Precast concrete walls should be laid-out differently if there is no access from the interior of the building. Swing (suspended) scaffolds are usually required to place the lines for the joints, which can often be snapped on the concrete or insulation. From these lines the connection hardware can be checked for location in the vertical and horizontal planes.

1.9.5 Complex Structures

Complex structures generally require a complete set of layout drawings for control and a complete tie-in to a master grid system. The complex structure could entail all the previously mentioned layout methods plus a higher degree of care and accuracy in the initial stages.

1.10 Lifting Devices and Handling Considerations

The precast engineer should locate lifting devices for ease of erection and connection of the precast concrete unit to the structure. Lifting points should be compatible with the method of shipping (flat or on edge) and be placed so that the rigging does not interfere with the structural frame during the precast installation operations. The same lifting

Fig. 1.10 (left) A special C-cradle and spreader beam arrangement allow these precast stadia units to be erected without the use of erection inserts; (right) C-hooks mounted on spreader beam used to erect architectural precast panel.

devices should be placed on the various precast concrete units so that frequent changes of the rigging can be avoided.

In general, the center of gravity of the unit should be computed and lifting points located to ensure that the precast concrete unit will hang level and plumb. If the center of gravity of the unit is not directly under the hook, the unit will not hang level.

If possible, the placing of lifting and handling devices should be planned so that little or no patching will be required after use. However, when the lifting and handling devices are located in finished edges or exposed surfaces, holes will require filling and patching. When this occurs, it may be necessary to obtain prior architect-engineer approval. Sometimes, specialized lifting equipment can be used to eliminate the necessity of patching exposed lifting and handling devices, Fig. 1.10.

The safety of erection personnel, the crane and the precast concrete units require that precast concrete lift points are not overloaded at any time during hoisting.

For non-typical sizes and shapes, the erector shall obtain exact information about how a unit is to be hoisted. It is the responsibility of the precaster to provide the erector with a hoisting diagram. The diagram should show the number of handling devices required for structurally safe handling.

During the rotation of precast concrete units with multiple hoisting lines, the center of gravity of the unit will shift and the loads on the hoisting attachments will vary. The precast engineer should evaluate the weight transfers during rotation of the member and design the hoisting points for the maximum load and variable direction of application of load.

The practice of placing one end of a panel on the ground during rotation to relieve the load on one or several of the hoisting points should be avoided. The hoisting points should be capable of supporting the precast concrete unit during panel rotation. It is not possible for a crane operator to determine or evaluate the correct amount of load that should be on the ground versus the amount supported by the crane.

DIVISION 2 — PRACTICES AND PROCEDURES

2.1 Transportation and Unloading

2.1.1 General

Efficient and orderly transportation of precast concrete products is a critical factor in a successful installation.

Prior to the start of erection and scheduling of delivery equipment, the jobsite shall be inspected to determine accessibility for the erection and delivery equipment.

At the preconstruction conference, arrangements should be made with the GC/CM regarding access into the jobsite, street use and traffic control, on-site storage areas and other factors affecting transportation. (See Sections 1.1 and 1.2)

Erection crews should be supplied with products and miscellaneous materials at a pre-established rate that will avoid delays at the erection site. Advanced planning and scheduling should consider details of loading at the plant, method of transportation, timing and sequencing of trailers at the jobsite, handling and erection, and storage at the jobsite, if required.

The latest revision of erection drawings shall be readily available at the jobsite.

2.1.2 Sequencing

The erection sequence is usually determined by the jobsite conditions and schedule (see Section 1.2). In addition, the sequence should take into account any special considerations pointed out on the drawings, special site conditions, and specialized equipment limitations. **In preparing the sequence of erection, the erector shall develop a plan to keep the structure stable and limit eccentric loading as much as possible. The erector shall take into consideration how and when the connections will be completed, weather limitations, and accessibility to subsequently hidden connections while preparing the sequence.**

Prior to commencing the work, the erector shall furnish the precaster with a list of specific pieces in order of their installation, as well as a list of unique loading requirements such as orientation of long members on the truck necessitated by site or rigging restrictions.

The precaster should ensure that all units are satisfactory for delivery before scheduling loading. The erector should provide ample time for the precaster to coordinate delivery of the members. This may necessitate lead time for final checking, possible cleaning and minor repairs resulting from damage or accumulation of dirt in storage.

It is recommended that the erector contact the precaster daily to coordinate the number of loads desired and the desired sequence and time of arrival of each specific load to the jobsite. Once a load arrives, the erector should unload it in a timely manner to allow transportation equipment to be utilized efficiently.

2.1.3 Access at the Jobsite

Costly project delays can be avoided if adequate access is provided on the jobsite. The GC/CM is responsible, unless specified otherwise and agreed upon by the subcontractors, for providing and maintaining adequate access into the site and operating space immediately adjacent to the building or in the building. Site conditions should allow erection and transportation equipment to proceed under their own power to a location where precast concrete members can be handled by the erection equipment directly from the transportation equipment into the structure.

Access for trucks and cranes including traffic control, street access, permits and unrestricted operations within the site shall be coordinated with the GC/CM, as required. Suitable access requires level, stable, well-compacted and maintained roadways or approaches and consideration for snow removal. Trees, shrubs, slabs, sidewalks, driveways and underground utilities and tanks and all items of a similar nature which need to be removed or protected should be clearly identified. **Overhead obstructions or lines which must be moved or de-energized and grounded shall be identified.**

Delivery trucks require jobsite roads that are firm, level and wide enough for the equipment involved to travel under their own power into position. Poor access conditions could result in mired and damaged delivery equipment with possible damage to product, such as torsional cracking of flanged members and even overturning of loads with a high center of gravity. This is also required for safe, efficient movement of cranes, Figs. 2.1.3(a) and (b).

Fig. 2.1.3(a) and (b) Good site access is critical for both crane and trucks.

Other site access considerations may include positioning the truck so that the crane can lift the unit from the truck and place it directly in its final location in the structure. Sometimes crane swing restrictions may require special access considerations. **Crane operating locations shall be firm and large enough to level the crane for safe, efficient operation. Ramps and inside access shall be provided as planned.** Some sites may require a qualified flagman during positioning of trucks.

It may be desirable to identify access routes and crane locations on the erection drawing to provide the GC/CM with a clear understanding of the access required to erect the precast concrete products.

2.1.4 Loading and Delivery

All precast concrete units should be delivered to the site clearly marked as indicated on the erection drawings. **Transportation and erection personnel shall obtain weights of the units from the precaster.**

Factors to be considered when delivering precast members from the plant to the jobsite are type, size, shape and weight of member, type of finish, weather, road conditions, method of transportation, type of vehicle, routing, distance to the job, and jobsite conditions. The shipper, not necessarily the precast concrete manufacturer or erector, is responsible for safe delivery of precast members. The precast concrete manufacturer and erector should advise the shipper of any special situations along the route and at the jobsite.

Units should be loaded and delivered:

1. to ensure safe transportation,
2. to meet various governmental transportation regulations,
3. with proper supports, frames, blocking, cushioning, and tie-downs to prevent or minimize in-transit damage. These materials are normally supplied by the precast concrete manufacturer and returned for repeat usage. All blocking, packing and protective materials should be of a type that will not cause damage, staining or other disfigurement of the units. The blocking points and orientation of the units on the shipping equipment should be as designated on the shop drawings.
4. to permit their removal from the load for erection in proper sequence and orientation to minimize handling,
5. to ensure orientation on the trailers and erection without unnecessary off-loading and rehandling, and
6. to ensure that sufficient product is delivered to the site in the prescheduled sequence to allow for orderly, efficient installation in the structure.

Federal, state, and local regulations may affect the size, weight and timing of shipping loads. When climatic conditions result in load restrictions, seasonal timing of the expected delivery should be considered. Limitations vary from one locale to another whether the shipment is by truck, barge, or rail. When large units are to be moved, a thorough check of local statutes is mandatory. This is usually done by the shipper. The shipper should take shipping restrictions into consideration when planning the route of travel and the delivery time to the site to insure uninterrupted delivery of product to the erection crew.

In addition to knowledge of legal requirements, it is necessary to study the route of travel to become aware of conditions including bridge restrictions, nearby buildings or power lines, restricted turning radii and natural obstacles likely to cause interference. Special attention should be given to roadway crowns which may affect the balance of the load and clearances under bridges. Part of the shipper's responsibility is to acquire the necessary permits to travel the selected route to the jobsite.

For special product shapes and sizes, the blocking points and orientation on the shipping equipment should be as shown on the shop drawings with bracing added to assure stability.

Flatbed or step deck trailers experience distortions or flexing during hauling. Dunnage between trailer deck and precast units should be sized to allow flexing of the deck without damage to the units. Typical transportation equipment is shown in Figs. 2.1.4(a) through (i).

Stacked members should be separated and supported by dunnage placed across the full width of each bearing location. Dunnage should be arranged in the same vertical plane and located where shown on the drawings, Figs. 2.1.4(j) and (k).

Units being transported should be securely tied-down to prevent overturning or sliding. Wherever possible, attachment to the vehicle should be made by fastening with a positive locking device, Figs. 2.1.4(l), (m) and (n).

The standard practice when tying a member to the trailer is to attach the tie-down device from the vehicle bed up and over the member and down to the vehicle bed on the other side at each dunnage support point. When using a chain or strap, it must be guarded to clear the corners or edges of the members to prevent damage, Fig. 2.1.4(o). The tie-down assemblies and securement devices should have an aggregate static breaking strength at least 1-1/2 times the weight of the precast concrete units being secured. An equivalent method is to have the aggregate static working load limit equal to at least 1/2 (0.5) times the weight of the precast concrete units being secured. Tie-downs need to meet the requirements of the FHWA and state transportation agencies.

The flat body of the trailer should always be swept clear of concrete particles or other debris. The driver should check load binders or other protectors for load security; being particularly certain the load cannot shift. The operator should be sure that any special instructions are followed. The operator should periodically stop to visually check the tractor/trailer rig and the load. Extra braking space should be allowed when hauling precast and prestressed concrete products. Due to the high center of gravity and large weight of some products, load shifting can occur when brakes are applied sharply. Loads with a high center of gravity also require a reduced safe speed when negotiating curves and corners. The shipper's responsibility for the product extends until the erection crew at the jobsite accepts the load.

For long slender units, anticipated deflection or vibration during transportation should be addressed with rigid bracing between units or by means of lateral stiffening. In extreme cases, strongbacks, stiffening trusses or frames may be required.

Care should be taken to minimize soiling, chipping, cracking or other damage to precast members during transportation. Corners ("L-shaped") and units with returns of unusual lengths are often shipped with bracing which should not be removed until just prior to erection. Spalling of edges or damage to returns should be minimized by proper attention to blocking and softeners, especially where tie-down chains or lifting slings are in contact with the member, see Fig. 2.1.4(o).

When GFRC panels are tied down to the trailer, it is suggested to use nylon straps rather than chains. Attention should be given to adequately support any top or bottom returns at the strapping points to avoid cracking of the GFRC skin, Fig. 2.1.4(p). Also, GFRC panels should be supported in a manner that minimizes relative movement between the stud frame and the skin.

GFRC panels should be loaded so that the GFRC skin does not support the weight of the unit. This is usually achieved by blocking between the trailer bed and the stud frame. To minimize cyclic loading of the skin anchors, it may be advisable to provide additional blocking to the skin, after the panel weight is supported on the stud frame. If panels are nested or stacked, consideration should be given to transfering the vertical load to prevent progressive crushing or other damage. This can be prevented by blocking the panels from stud frame to stud frame. Specially designed support frames may also be used to prevent damage from occurring.

The forces imposed on a member during transportation are affected by orientation of the member, location of the supports and binders, and condition of the transporting vehicle, roads and site. Sometimes two or more units can be transported together, side by side, and tied together to provide the necessary lateral strength.

The following are the manufacturer's and shipper's responsibilities:

1. The manufacturer must decide which transportation method to employ. Therefore, it is also their responsibility to anticipate and verify behavior of the precast concrete member during loading and transportation.
2. The shipper is responsible for providing competent drivers who will exercise reasonable judgement to protect the precast concrete members during transportation.
3. The shipper is responsible for the timely delivery of the precast units in proper sequence. This is provided that the loads are accepted in sequence and that the anticipated delivery times are adequate for the project conditions and are agreed to prior to commencement of deliveries. Likewise, the erector should unload the delivery vehicle as soon as conditions permit so as not to delay the shipper unnecessarily. Delivery of product should include a reasonable unloading time allowance.

(a) Tilt frame trailer unloaded.

(b) Tilt frame trailer loaded.

(c) A-frame on single drop low boy trailer.

(d) Slant frame on single drop trailer.

(e) A-frame on flatbed trailer.

(f) A-frame on double drop trailer.

(g) Nine-axle steerable hauling unit.

(h) Twelve-axle remote steerable hauling unit.

(i) These finger racks allow the precast spandrel beams to be transported in the vertical position. Note the adjustable screw clamps wiich secure the beams to the rack.

Fig. 2.1.4(a) through (i) Typical transportation equipment.

Fig. 2.1.4(j) A typical load of double tees on a flat bed trailer. Note vertical alignment of dunnage.

Fig. 2.1.4(k) A specialized double tee load on an expandable trailer with a rocker frame.

Fig. 2.1.4(l) Use of chain and binder.

Fig. 2.1.4(m) Use of fabric straps.

Fig. 2.1.4(n) Tilt support to haul 15-ft wide double tee.

Fig. 2.1.4(o) Member with corner blocking and softeners.

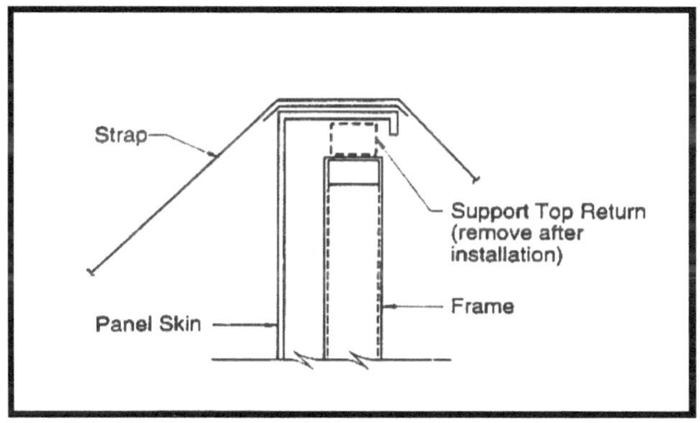

Fig. 2.1.4(p) Recommended strapping procedure for GFRC panels.

2.1.5 Unloading

The erector should develop, together with the precaster and shipper, clearly defined product acceptance criteria and procedures. This should help to avoid later disagreement concerning product damage upon arrival at the jobsite.

A visual inspection of the units shall be made on the trailer prior to unloading. This will allow for a close view of the product so that defects will be more easily detected. The following are items that should be checked:

1. Identification: Check that the quantity and mark numbers agree with shipping tickets.

2. Cracks: Check for cracks on all surfaces.

3. Camber: Prestressed products such as double tees, beams and plank normally have a noticeable camber. Check for negative or abnormal camber. **Prestressed products with negative camber shall be brought to the attention of the precast engineer immediately.**

4. Spalls: Check for spalls that may have occurred during shipping.

5. Scratches: Inspect for scratches on architectural products.

6. Red Tags: If a product has a red tag attached, it requires repairs that may not have been completed.

7. Lifting Loops and Inserts: Check to assure that lifting devices are in good condition, correct type and that threaded inserts have at least the minimum thread depth.

In the event that any irregularities are detected on the product or the structural integrity appears to be impaired, the erector shall immediately notify the precaster so that a decision can be made on product acceptance or rejection.

The precast concrete manufacturer's instructions for handling shall be followed. All precast concrete units shall be handled in a position consistent with their shape and design by means of approved devices which are located and identified on the erection drawings. Units of irregular shape should have support and handling points clearly designated. **Prestressed products shall be maintained in an upright position and lifted and supported at points near the ends unless erection drawings designate otherwise.**

Prior to unloading the unit from the transport vehicle, all bracing, chains, straps, and protection for edges of the units shall be carefully removed. Chains, binders and straps shall not be removed from all the units unless stability of the load is assured. If belts or slings are used in unloading, protective materials should be used at points of contact with units to reduce damage. **Multiple units shall not be handled simultaneously unless palletized or designed to be moved in this manner.**

To safely unload precast concrete members, the truck and trailer should be on firm, level ground. Depending on conditions, the trailer should be blocked and/or the trailer landing gear lowered. **Employees shall stand clear of precast concrete members while they are being unloaded and moved.**

The units shall remain under control when hoisting and during placement. Control may require taglines. Taglines should be used on the units when necessary to control the load to prevent hitting personnel, the crane boom, fixed objects or electrical wires, Fig. 2.1.5. Taglines should be properly controlled to prevent snagging.

Fig. 2.1.5 Use taglines when necessary to control load.

During erection, wind and or other conditions may cause difficulties in controlling wall panels or other units with large surface areas, and may induce lateral and torsional stresses in the crane boom. Therefore, tag lines on the precast units are often necessary to control the load to prevent it from uncontrolled swinging and hitting the crane boom, fixed objects or electrical wires.

Frequently, panels on one side of an A-frame trailer should be rechained or otherwise secured with come-a-longs before unloading the other side. Again depending on the method the shipper used to brace the unit during transportation, the erector may need to place an initial load on lift lines with the crane before removing all of the chains or other devices holding the piece in place.

Units should be lifted slowly from the transport vehicle. If binding occurs, the units should be lowered and the obstacle restraining the unit removed. **The exterior unit shall always be unloaded first from a stack to prevent chipping and scraping; units shall not be slid out from the middle of a stack.** Whenever possible, on a load with many units, balance of the trailer should be maintained by unloading alternate sides of the vehicle or by blocking the vehicle to prevent tipping. **Evaluation shall be made for unbalanced and unstable loads during the sequence of removing units from the trailer.**

Controlled rotation of units shall be accomplished in the air and not by turning the units on the truck or ground, unless the erection drawings or precast engineer indicates the units are properly designed to resist such handling. Also, see Section 2.2.8 for discussion.

Methods of unloading shall avoid soiling, cracking, chipping, and other damage to precast units. Edge protectors should be used to prevent damage from direct contact between wire rope slings and concrete units, especially those with fragile shapes or projections. Care should also be taken, when slinging or landing units, to prevent corners and corbels from striking other units or becoming damaged by unintentional point contact or impact loading.

2.1.6 Site Storage

It may be desirable to maintain an inventory of precast products on the project to prevent delays in erection. This can be achieved by storing several loaded trailers on the jobsite. Leaving the precast concrete members on the trailer is often the most desirable solution for on-site storage. The trailer usually provides clean and safe storage and eliminates extra handling and possible damage. If a loaded trailer is stored (dropped) at the jobsite, the shipper should carefully stabilize the trailer to prevent tilting or overturning. This is especially true on sites which are frozen and subject to thawing. **The stored trailer shall be parked on firm, level ground. Sufficient dunnage shall be placed under the landing gear dollies to prevent them from sinking. This shall be the responsibility of the erector only if the erector is responsible for transportation.**

Site storage other than on trailers should be avoided, as it is expensive and can result in damage to the product. The desirable method of precast concrete erection is the unloading of members directly from the delivery trailer to their final location on the structure without storing them on the jobsite. If a piece must be shipped out of sequence, the precaster should announce that a piece is coming out of order so that the erector can plan for it. The erector has a few options:

1. The member can be installed in its correct location if the crane can reach it and if it does not interfere with installation other units.

2. It can be stored temporarily on the ground, but this is often not desirable or feasible due to site constraints.

3. The piece can be left on the trailer or loaded onto another trailer.

Often, the second option is the only one available to the erector. Most jobsites are congested and finding a location out of the way that is clear of debris and level enough to allow safe storage is often very difficult. **When storing products on the ground, care shall be exercised in determining that the dunnage will support the unit without settling, warping, overturning or damage.** Softeners are always required both on the dunnage and at the points where a panel leans against a structure. If a panel is tall, it should be tied back to prevent overturning. **If a panel is stored against the building, care shall be taken to prevent damage to the building or the unit.** The product should be protected against soiling or damage by one of the other trades on the project.

If on-site storage is for more than a few pieces, storage areas should be provided by the GC/CM and should be level and firm, well-drained, and located where there is little chance of damage due to other construction activity. **Any temporary storage at the jobsite shall be carefully planned to avoid overstressing the units.** The area may have to be stabilized to prevent differential settlement or twisting of members. This is especially important for members stored by stacking. Units should not be stored on frozen ground without proper safeguards to prevent settlement when the ground thaws.

Stacked units temporarily stored in an on-site storage area shall be separated and supported at blocking points designated on the erection drawings, Fig. 2.1.6. Dunnage shall be arranged in vertical planes and deep enough to provide clearance for lifting devices. Also, units should be stored with identification marks visible.

When stacking stemmed members or plank, dunnage shall not be continuous over more than one stack of precast units. Stacking of members shall be such that lifting devices will be accessible and undamaged. The upper members of a stacked tier shall not be used as additional storage area for shorter members or equipment.

The erector is responsible for proper on-site storage, whether planned or unplanned. After the erector inspects the material prior to unloading and accepts the material from the shipper, responsibility is assumed for general safekeeping until the precast units are installed.

Fig. 2.1.6 These precast double tees are properly stored on sound dunnage at their lifting points.

2.2 Rigging

2.2.1 Training Programs

An erector that employs personnel who do rigging shall develop training programs, including means of testing, to assure that the personnel are competent to perform the operations.[1] The training programs shall include, but not be limited to:

1. **The use of ropes, shackles, hooks, snatch blocks, wire ropes, chains and fabric slings, timbers, hoisting principles, rollers, scaffolds, weight estimations, center-of-gravity, factors of safety, and the force-effect of angular lines in load lifting.**
2. **Safe attachment of slings for straight lifts, basket hitches, chokers, and multiple-bridle lifting.**
3. **Safe placement of sling hooks.**
4. **Hook safety latches.**
5. **Load-sling adjustments to keep the center of gravity in line with hook pull.**
6. **Risks and precautions when rigging near overhead power transmission lines.**
7. **Use of spreader bars for lifts and adjustment to keep the center of gravity in correct alignment for balanced lifts.**
8. **Inspecting rigging equipment.**

2.2.2 Rigging Selection

After selecting hoisting equipment, properly sized rigging shall be selected to permit the precast concrete units to be lifted and installed safely.

Wire rope slings are usually the major component of the rigging assembly; on occasion web belts or chain mesh belts may be required.[2,3] Hooks, shackles, rolling blocks, closed links and lifting plates can be used with the slings to complete the system, Fig. 2.2.2. All of these items are commercially available with test data to substantiate the safe working load for the available sizes of each product. Prior to utilizing any of these items in a rigging assembly, it is imperative that the user consult the manufacturer's published charts and information in order to be aware of the conditions and limitations for which the safe working load data applies.

If rigging assembly components are custom made for a particular situation, a competent person or professional engineer shall design such components with full knowledge of their range of intended use. Before a custom made lifting apparatus is utilized as standard equipment, its capacity shall be clearly labeled for future reference.

It is necessary that the erector receive accurate, individual unit weights and dimensions, and if applicable, special handling requirements from the precast concrete manufacturer. The weight, handling requirements and size of the unit to be lifted are the critical factors in rigging selection. **With this information, the erector shall determine the rigging requirements.**

Fig. 2.2.2 Attaching rolling blocks and slings to trip panel.

Fig. 2.2.3 Load blocks compatible with load line diameter.

2.2.3 Load Block Selection

Another decision which shall be made for the rigging is the capacity of the load block to be used with the crane. The criteria for load block selection are

1. The heaviest lift to be made,
2. The number of parts of load line, and, hence,
3. The minimum number of sheaves in the block required to make the lift.

The hoisting equipment supplier will generally supply a load block compatible with the crane's capacity and load line diameter. The final determination is the number of the line parts to be reeved into that block, Fig. 2.2.3. To make this determination, divide the weight of the heaviest lift by the safe working load of a single part of load line. Round that answer up to the next whole number to determine the *minimum* number of line parts. The safe working load of a single part of load line is listed on the manufacturer's load chart for the hoisting equipment.

2.2.4 Sling and Spreader Selection

Other items to be selected are the wire rope slings and spreaders. The most commonly used wire rope for rigging slings is of IPS-IWRC 6 x 19* construction with flemish eyes and pressed metal sleeves, Table 2.2.4(a). The length of the slings is a function of the size of the member to be erected and the diameter is a function of the weight of that member. Slings and spreaders should be of adequate length so that the lifting devices are not overstressed.

The simplest method to determine sling length is to measure the maximum distance between lifting devices on the longest member to be erected. In the case of a 2-point pick, select slings each of which are of a length equal to that distance. Using 2 such slings, an approximate equilateral triangle is formed with the member with an included angle, Ø, of 60 deg. between sling and member. Two-crane lifting may be desirable for long-span members to eliminate the need for long slings and to achieve better control when placing a member.

To determine the proper wire rope diameter, refer to a wire rope capacity chart available from the suppliers. Under the column for 60 deg. included angle for the above example, find a rated capacity which most nearly exceeds the weight to be lifted and read across to determine the required diameter, Table 2.2.4(a).

Care shall be exercised when using capacity charts. This is because the included angle listings generally assume that two slings are being utilized. Hence, the weight to be lifted is used directly to determine the diame-

* Improved Plow Steel — Independent Wire Rope Core with 6 strands of 19 wires each.[10]

(a) Shackles attached to lifting hardware.

(b) Hook attached to lifting hardware. (Note: use of chain come-along to level member)

(c) Shackle and hook attached to lifting hardware.

Fig. 2.2.4 Attaching slings to precast members.

WIRE ROPE SLINGS 6 x 19 Classification Group, Improved Plow Steel, IWRC						
	MAXIMUM SAFE WORKING LOADS — POUNDS (Safety Factor = 5)					
Rope Diameter (Inches)	Single Vertical Hitch 	Single Choker Hitch 	Single Basket Hitch (Vertical (Legs) 	2-Leg Bridle Hitch & Single Basket Hitch With Legs Inclined  		
				60°	45°	30°
3/16	600	480	1,300	1,100	900	650
1/4	1,150	860	2,300	2,000	1,600	1,150
5/16	1,750	1,300	3,500	3,000	2,500	1,750
3/8	2,550	1,900	5,100	4,400	3,600	2,550
7/16	3,450	2,600	6,900	6,000	4,900	3,450
1/2	4,700	3,500	9,400	8,150	6,650	4,700
9/16	5,700	4,200	11,400	9,900	8,050	5,700
5/8	7,100	5,300	14,200	12,300	10,000	7,100
3/4	10,200	7,650	20,400	17,700	14,400	10,200
7/8	13,750	10,300	27,500	23,800	19,400	13,750
1	17,950	13,450	35,900	31,100	25,400	17,950
1 1/8	22,750	17,000	45,500	39,400	32,200	22,750
1 1/4	28,200	21,200	56,400	48,800	39,900	28,200
1 3/8	34,800	26,100	69,600	60,300	49,200	34,800
1 1/2	41,300	31,000	82,600	71,500	58,400	41,300
1 5/8	48,600	36,400	97,200	84,200	68,700	48,600
1 3/4	55,900	41,900	111,800	96,800	79,000	55,900
1 7/8	65,400	49,000	130,800	113,300	92,500	65,400
2	72,600	54,500	145,200	125,700	102,700	72,600
2 1/4	90,300	67,600	180,600	156,400	127,700	90,300
2 1/2	111,800	83,700	223,600	193,600	158,100	111,800
2 3/4	131,100	98,200	262,200	227,000	185,400	131,100
				If used with Choker Hitch multiply above values by 3/4. 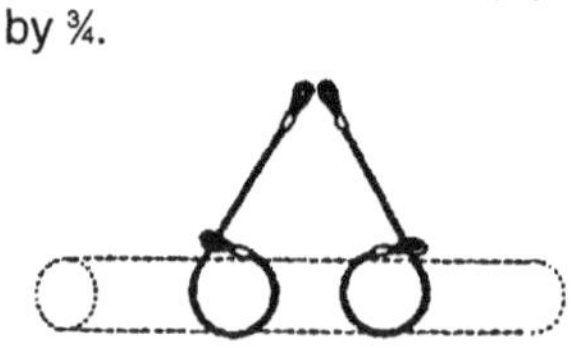 For Double Basket Hitch multiply above values by 2. 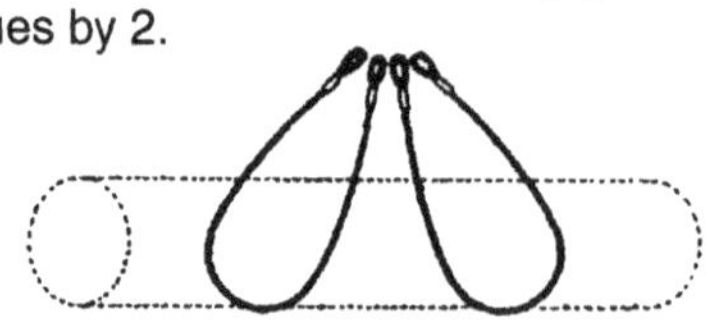		
Note: Table values are for slings with eyes and thimbles in both ends, Flemish Spliced Eyes and mechanical sleeves.						

Table 2.2.4(a) Rated capacity of wire rope slings.

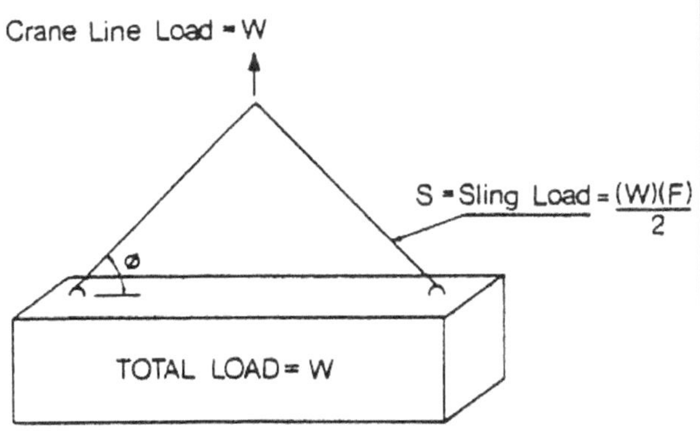

	Multiplication Factor "F" for the total load on the hoisting sling with an inclination = Ø					
Ø	90°	75°	60°	45°	30°*	15°*
F	1.00	1.04	1.16	1.41	2.00	3.86

Note: Ø usually not less than 60° to 70°
*Not recommended

Table 2.2.4(b) Determination of force in inclined lift lines.

ter of both slings. **Even though these charts include a substantial safety factor, usually 5, the capacities shall not be exceeded.**

The length of the eye shall be specified to be of sufficient size to permit it to slip easily onto the hook of the load block and move freely on the hook.

To attach the slings to the precast member, properly sized hooks are usually attached to the sling eyes with properly sized shackles, Figs. 2.2.4(a), (b) and (c). When sizing these two items, it must be remembered that the actual pulling force on the hook and shackle increases as the angle, Ø, between the sling and member decreases, see Table 2.2.4(b). For example, when the slings and their accessories are used with an angle of 30 deg., the force on the slings is twice as high as if the slings were lifting vertically. **The rated capacity of the individual hook and shackle, usually embossed on the item, shall be compatible with the sling used and shall exceed the load to be lifted, considering the angle between the sling and member.**

For erection of precast units, the sling angle shall never be less than 45 deg. from the horizontal unless the effect of the horizontal component of the lifting force on the member is checked against member and lifting device capacity by the precast engineer.

The lifting anchors which are cast into precast products generally are not designed to withstand the excessive lateral (shear) loads which result from slings pulling at an angle less than 45 deg. The most commonly used sling configuration results in the 60 deg. angle discussed previously. If, for headroom considerations, a short hookup is required, a spreader beam properly designed for the load, should be utilized so that the capacity of the insert is not exceeded.

2.2.5 Lifting Anchors and Hardware

Lifting anchors cast in precast members vary depending on the member's use in the structure, the size and shape of the member, and the precast concrete manufacturer's preference. A common device is an anchored loop of aircraft cable (Table 2.2.5), cold-drawn wire or prestressing strand. The hook placed through the loop for hoisting must be of sufficient capacity for the load yet be able to slip easily into the loop. If this is not possible, then a properly sized shackle must be used rather than a hook.

The erector should verify that (1) loops are of equal projection when used as a pair or in groups so that the unit will hang level and plumb, (2) bundled or multiple strand loops are positioned so that they equally share the load, and (3) individual wires are not broken or damaged. **If these conditions occur the erector shall notify the precaster.**

2.2.6 Engineered Lifting Devices

There are many proprietary lifting systems available. A lifting system consists of an anchor cast into the concrete and lifting hardware that is designed to couple specifically with that anchor. The hardware usually has a ring or bail to which the rigging is attached. Some systems have different anchors to accomodate specific loading conditions and are available in different capacities. Some anchors are not recommended to accomodate all directions of lifting from shear to tension. This may be of particular concern if the anchor is used as a side lifter or through various angles if a panel is being rotated. Lifting anchors should be capable of supporting at least four times the maximum load throughout all directions of the lift. The lifting hardware should have a five-to-one safety factor.

Diameter (in.)	Capacity (kips)[b]
3/8	3.6
7/16	4.4
1/2	5.7

a. 7 strands with 19 wires each.

b. Based on a single strand with a factor of safety of 4 applied to the minimum breaking strength of galvanized cable. The user should consider embedment, loop diameter and other factors discussed for strand loops. Aircraft cables are usually wrapped around reinforcement in the precast product.

Table 2.2.5 Safe working load of 7 x 10 aircraft cable used as lifting loops. *Contact precaster for capacity of strand lifting loops.*

Another lifting device is a swiveling D-ring plate bolted to a threaded anchor. **The swivel and bolt diameter shall be the same size, Fig. 2.2.6(a).**

Although coil-threaded anchors have been commonly used in the past, they are discouraged when other types of anchors are able to be used. Insufficient thread engagement is the most common problem with this system and can be dangerous if the bolt is not fully threaded into the anchor. **When using coil-threaded anchors, the following precautions shall be followed:**

1. **It is imperative that thread engagement be not less than 2-1/2 times the nominal bolt diameter. Because anchors are sometimes recessed, the erector shall locate the actual top of the steel anchor, as some threads may be concrete cover and not steel. If in doubt, the erector shall scratch the threads with a metal object to verify anchor location.**
2. **Caution shall be used in tightening bolts as they may break out the opposite face of a thin panel or may strip the insert threads.**
3. This type of anchor is particularly affected by foreign materials such as sandblast grit and/or ice which may have accumulated in the insert. **Care shall be taken to remove foreign materials from inserts to ensure full thread engagement.**
4. **A minimum of two coil inserts shall be used to lift a unit so that the unit or lifting line cannot spin causing the bolt to unscrew from the insert.**
5. **Bolts of proper length shall be used to prevent extensive shimming.** Bolts that are too long may bend and fail.

The precast concrete manufacturer should be consulted regarding the design intent of all erection inserts. Connection hardware should not be used for lifting or handling, unless specifically designed and designated for this purpose on the erection drawings.

Some long and/or heavy precast members may have sleeves through the member or anchors with swivel hardware for use in attaching erection hardware, Fig. 2.2.6(b).

There are situations when it is necessary to develop special lifting systems. **If so, the system shall be designed by a professional engineer and tested to verify that the anchor will exceed a safety factor of four-to-one and the hardware five-to-one.**

Fig. 2.2.6(a) D-ring swivel used with threaded anchor.

Fig. 2.2.6(b) Column being rotated with through-pin.

*2.2.7 Two- and Four-Point Picks**

The simplest precast members to lift are fascia panels or beams. These units generally require a 2-point pick. Two properly sized slings are attached to the lifting hook of the hoisting equipment. At the far end of each sling, a safety hook or shackle is attached to the sling eye. The hook is attached to the lifting hardware compatible with the lifting anchor provided by the precast concrete manufacturer.

(a) Typical 4-point pick for double tee.

(b) Typical 4-point pick for long spandrel beam.

Fig. 2.2.7 Four-point picks.

A rolling block (snatch block) can also be used to effect a 2-point pick. The rolling block is shackled to the hoisting hook and a single long sling is threaded over the block. Hooks and shackles are attached to each eye of the sling for attachment to the precast unit. As the load is lifted, the single sling forms a 2-legged spreader set about the rolling block. **This rolling block method shall be used only when the vertical length of the precast member exceeds the horizontal width by at least 25 percent, to prevent the sling from rendering through the block.**

The advantage in using a rolling block for the 2-point pick is that if the unit hangs out of level, a come-a-long can be attached to the hoisting hook and to the low side erection device of the member. The come-a-long is then able to level the unit while the sling rolls through the block.

If the unit to be erected is long and/or heavy, such as a spandrel beam or double tee, a 4-point pick may be desirable or necessary, Figs. 2.2.7(a) and (b). Four equal length slings, similar to the 2-point pick, can be used. Alternately, two long slings can be used with a matched pair of shorter slings shackled to the end of the longer slings to form a "Y", Fig. 2.2.7(b).

2.2.8 Rotation of Members

If the unit to be erected requires rotation from its horizontal delivery orientation to a final vertical orientation, such as for a tall, slender precast wall panel, two hoisting methods are commonly used. If a crane with two hoist drums is used, the main load line is connected to the top of the unit using a single sling and rolling block. The auxiliary load line is attached to an erection insert cast into the edge or back of the member, depending on the delivery orientation, near the bottom, Fig. 2.2.8(a) through (f).

The load lines should be located approximately over the center of gravity of the unit as it rests on the delivery vehicle. Each load line is lifted alternately until the unit has cleared its dunnage blocks or A-frame. If this is not done carefully, the unit will drift as it leaves its supports.

The unit is then swung clear and the main load line is raised slowly while the auxiliary load line is held stationary. As the unit is raised, the upper sling passes through the rolling block, allowing the unit to turn to its final, vertical orientation.

It is recommended that when this method is used, the unit be placed before the crane so that an imaginary line between the top and bottom inserts is parallel to the crane

* For ease of use by the erector, this Manual uses the actual number of lifting devices in the products to identify the lifting points, i.e., 4 point pick has 4 lifting devices. To account for structural behavior, industry practice refers to a line of support or resistance. Therefore, the previous 4 point pick would be referred to as a 2 point pick by the structural engineer. Caution should be exercised by erector when discussing lifting points versus number of lifting points with the structural engineer

(a) 6-point rotating.

(b) 4-point rotating.

(c) 8-point rotating.

(d) 4-point flat rotation using edge picks.

(e) 4-point rotating on edge.

(f) 3-point rotating on edge.

Fig. 2.2.8 Various methods of rotating (tripping) wall panels from their delivered to erected position.

Fig. 2.2.8(g) Two conventional crawler cranes are used to rotate a long double tee wall panel.

boom when the unit is hoisted. If this cannot be done, then the longest crane boom which is practical should be used (the minimum boom length should be 5 times the distance between the picking points of the two load lines). If a short boom is used and the member is perpendicular to the crane boom, the included angle between the main load line and auxiliary load line can be large, depending on the length of the unit. If that angle is large, one of the load lines could possibly jump out of its sheave at the crane boom tip. Further, the lateral forces induced by such a large included angle will tend to orient the member to a position parallel to the crane boom as soon as it clears its dunnage points. This natural rotation is difficult to control, particularly for heavy members. Care must be taken to anticipate and control this movement to avoid injury or damage. A tag line should be used to hold the unit while it is being rotated so that it does not swing uncontrollably.

Two cranes can also be utilized to rotate members from horizontal to vertical. The first crane's load line is attached to the top of the member using a sling and rolling block. The second crane's line is attached to the edge or back inserts and each crane lifts alternately until the unit has cleared its dunnage blocks or A-frame. The unit is then swung clear and the first crane raises its end until the member is vertical. The second crane's load line is then slackened and can be disconnected, Fig. 2.2.8(g).

To rotate relatively short, light units with a single load line, alternative methods are available. One is to outfit the load hook with a sling and rolling block for attachment to the top erection inserts and a chain fall for attachment to the edge or back inserts near the bottom. Effectively, a 2-point pick is being made; the sling/rolling block being one leg and the chain fall being the other. The chain fall should be adjusted to assure that the unit remains level when lifted. After the unit has been hoisted and swung clear, the chain fall can be lowered to allow the unit to rotate to its vertical position. **The chain fall shall have proper capacity and sufficient chain overhaul to allow the unit to rotate its full 90 deg.**

Another method for rotating units, if only one load line is available, is the turning cradle. This is a custom-fabricated device resembling a teeter-totter, on which a member is placed in its horizontal orientation, providing support along the full length of the member. The hoist hook is then attached to the top of the member with the appropriate rigging. As the unit is raised, the member is rotated to a near vertical position about the axis of the cradle. It is then lifted from the cradle and installed, Fig. 2.2.8(h).

Fig. 2.2.8(h) Custom-built turning cradles allow panels to be rotated using a single load line from the crane.

Fig. 2.2.8(i) Use of strongback and brace on long slender members.

The turning cradle shall provide structural support for the member to be turned without marring or damage. The cradle shall be designed not to rotate while being loaded. The designer of the precast member to be rotated should be consulted to assure that the use of the turning cradle does not introduce unanticipated forces into the member.

For long slender members which are not prestressed, it may be necessary to use a strongback or external bracing in order to reduce tensile stresses in the member, Fig. 2.2.8(i).

While new commercial rigging devices are continually being developed, and custom-made devices are limited only by the erector's ingenuity, it is important at all times to work within the design limits of the individual rigging components. Conservatively designing the rigging, while perhaps somewhat more costly, helps to assure safe, and ultimately, economical construction.

2.2.9 Lifting with Two Cranes

Lifting a load with two cranes should be avoided unless dimensions, weight of load, or characteristics of the lift prevent use of a single crane. The operation requires careful planning and an accurate assessment of the share of the load carried by each crane. There should be a full appreciation of how and to what extent this can vary should the load come out of level, or should one or both cranes have to boom-up, boom-down, travel or swing resulting in the hoist lines coming out of plumb. If any of these circumstances occur, an additional load can be introduced to either or both cranes which may affect crane stability or cause structural damage to the crane boom.

Special lifting tackle, such as a spreader bar or equalizer beam, may be necessary to suit the maximum variation in load distribution and direction of application which can occur during tandem lifting.

If possible, cranes of equal capacity and similar characteristics should be used. Each crane and the lifting tackle should have a capacity margin of 25% greater than needed for its calculated share of the load, Fig. 2.2.9.

Fig. 2.2.9 Two-crane tandem lift.

When performing a two-crane lift, it shall be under the direction of a competent person even though two signalers may be required. With any two-crane lift, it is imperative that both crane operators have a complete understanding of what the other is going to do. Also, care must be exercised when the cranes are set up so that both stay within their capabilities, but do not interfere with each other, throughout rotational maneuvers.

2.2.10 Rigging of Typical Precast Units

Rigging arrangements, devices and picking points are shown in Figs. 2.2.10(a) through (f) for different precast units and for either straight lifts or rotation in the air. The designer of the units should determine lifting stresses, select proper locations for picking points and provide all necessary information and details on shop or erection drawings.

DESCRIPTION

This system enables the erector to control the load all the time with the top line because the center of gravity of the precast member is above the center of the rolling block. To ensure that this condition does not change during rotation the following stability condition has to be met:

$$e > \frac{\ell}{2} - \frac{\sqrt{\ell^2 - b^2}}{2} - a$$

This minimum rolling block length is important to prevent a sudden shifting of the load while it is being rotated.

COMMENTS

1. Stresses during erection are reduced because of 3-point lifting.
2. Member should be taken directly from the truck to avoid double handling.
3. Method is not good when top-in-form haunches or top projecting dowels exist.
4. Both lines of the crane are required to carry almost the total load of the member (line 2 carries 90-95% at the beginning and line 1 carries 100% at the end).
5. Proper length cable and rolling block is required.
6. Stripping stressed may require additional lifting loops since top cantilever may be too long in some cases.
7. Range of use depends on cross sectional properties of the member but usually is 45-90 ft. (7-50 tons).
8. Minimum boom length required is longer than for other methods.
9. Critical area for stresses is usually top cantilever.

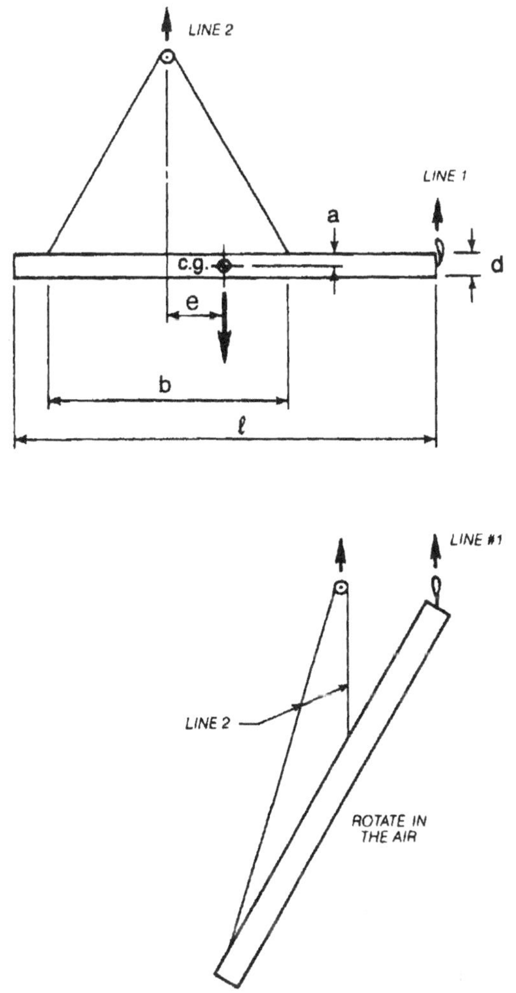

Fig. 2.2.10(a) Two-line lifting with rolling block — top heavy.

COMMENTS

1. Member may be taken directly from the truck, or rotated in the air using line 2.
2. When member has been designed to be handled with two lines, care must be taken to lift first with the bottom line and then rotate in the air.
3. Range of use depends on the cross sectional properties of the member but usually is limited to 30 ft. using top lift loop only or 20-60 ft. with top and bottom lift loop.

RECOMMENDED FOR UNIFORM SECTION MEMBER

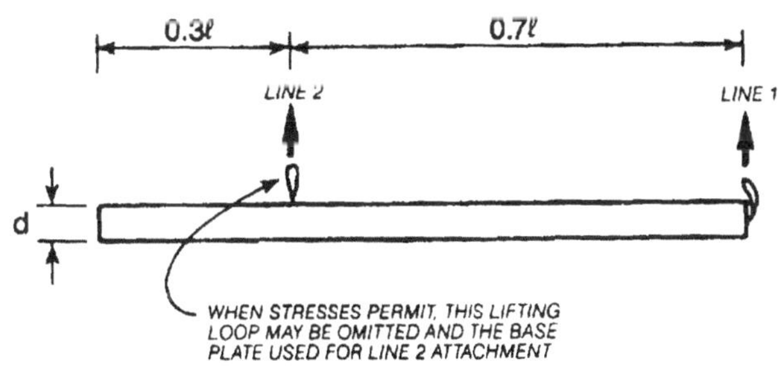

Fig. 2.2.10(b) Two-lifting loops or base plate and top-lifting loop.

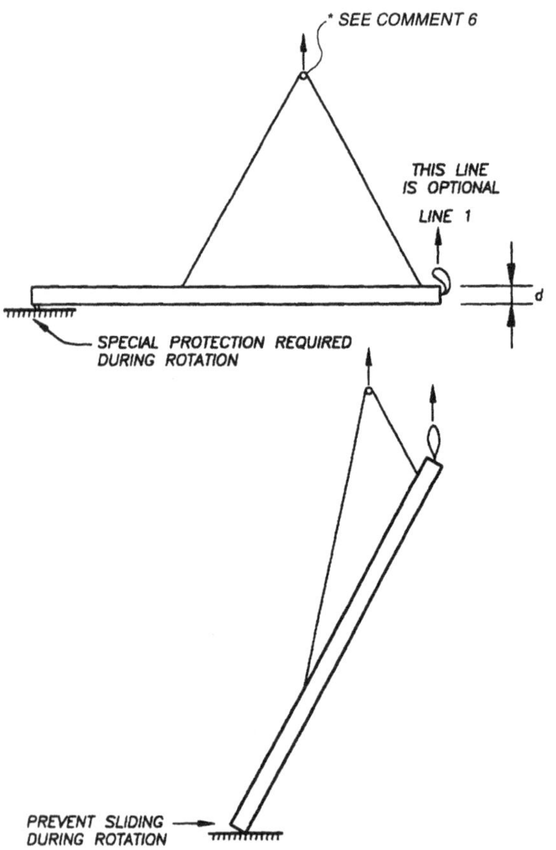

COMMENTS

1. Bottom of member needs to be protected during rotation.
2. Double handling usually required since it is not recommended to rotate on truck.
3. Top line may be omitted when final adjustment is not a problem because member is not perfectly vertical.
4. Method is not good when top-in-form haunches or top projecting dowels exist.
5. Range of use depends on the cross sectional properties of the member but usually is 45-90 ft.

*6. For this system to work without top (optional) line, line 2 must be above the center of gravity.

Fig. 2.2.10(c) Rolling block — balanced or bottom heavy.

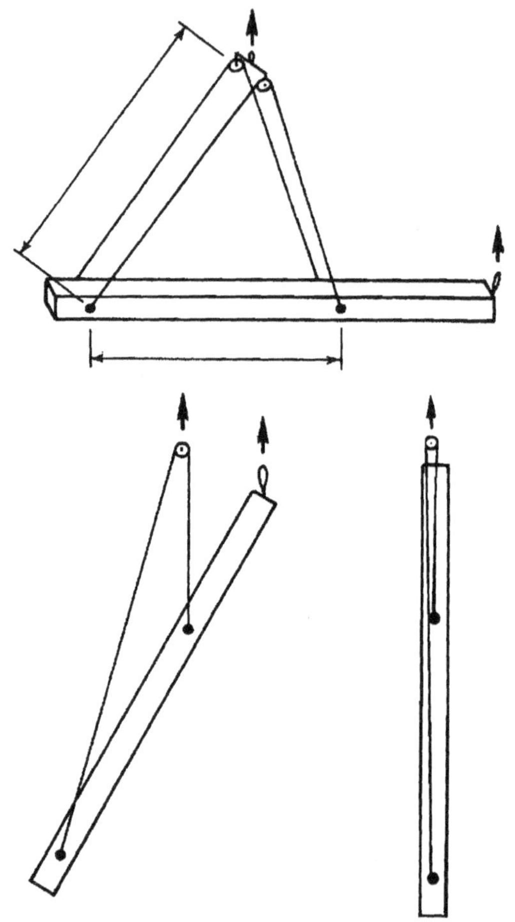

COMMENTS

1. Special rigging required.
2. Additional patching work required since sleeves are in addition to lifting loops or inserts required for stripping.
3. Only one line required.
4. System is very good to prevent interference with top or top-in-form haunches or dowels.
5. Double handling required since member has to be taken from the truck before rotation.
6. Stresses in the concrete are lower than with other systems.
7. Length of boom required is shorter than with most systems.
8. Range of use depends on the cross sectional properties of the member but usually is 45-90 ft.
9. Slings must be long enough so that spreader bar clears top in vertical position.
10. See Fig. 2.2.10(c).

Fig. 2.2.10(d) Top and bottom sleeves — bottom heavy.

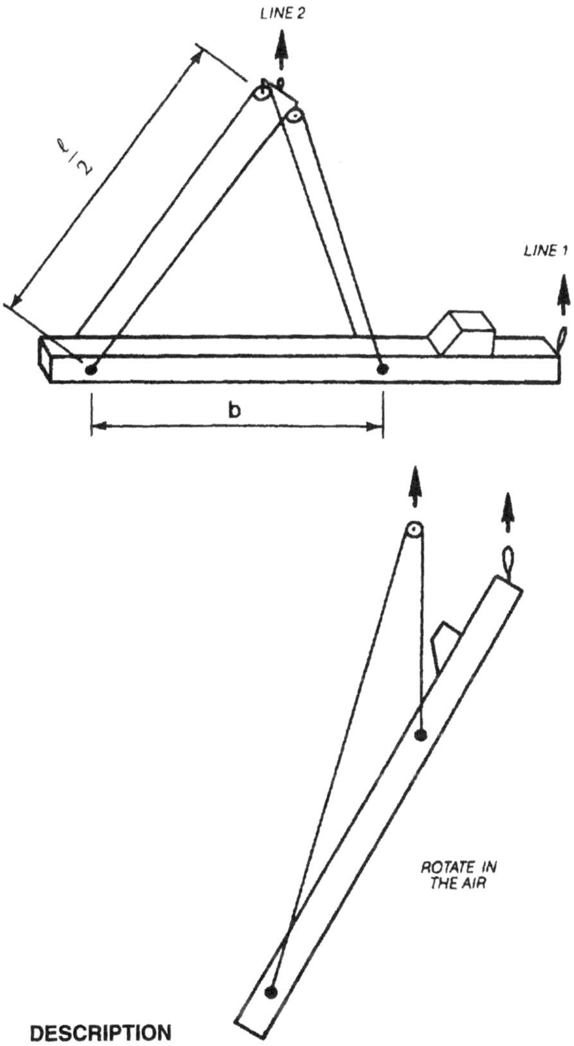

DESCRIPTION

This system provides control of the load all the time with the top line, provided that the stability condition below is checked.

$$e > \frac{\ell}{2} - \frac{\sqrt{\ell^2 - b^2}}{2}$$

See Fig. 2.2.10(a)

COMMENTS

1. If required member can be taken directly from the truck.
2. No special protections required.
3. System is very good to prevent interference with top-in-form haunches or dowels.
4. Top lifting loops or inserts are required for stripping.
5. Only line 1 of the crane has to carry the total load.
6. Minimum boom length required is longer than for other methods.
7. Normal range of use is for members 45-100 ft.

Fig. 2.2.10(e) Sleeves — top heavy.

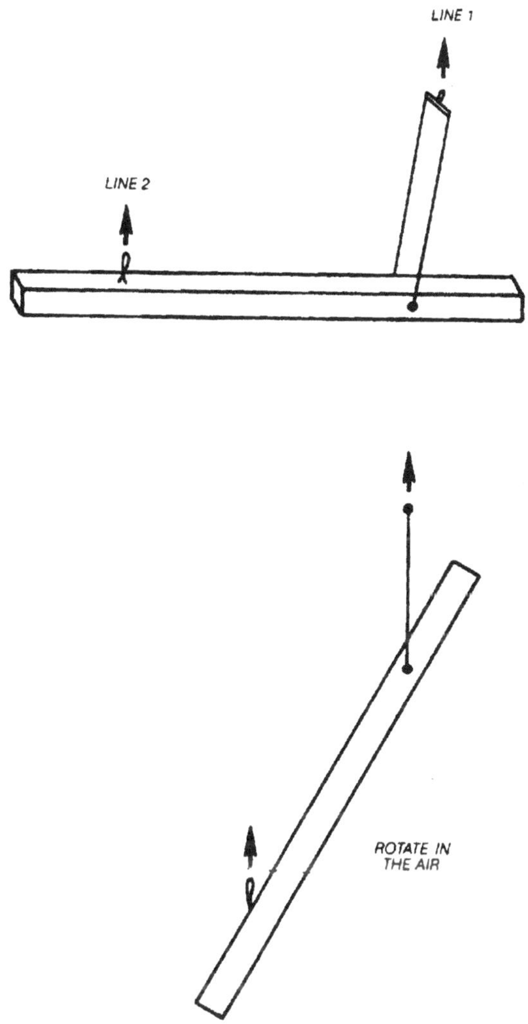

COMMENTS

1. System gives flexibility to locate lifting loop and sleeves.
2. Top sleeve avoids interference problems with haunches or dowels projecting at the top or upper surface of member.
3. Only line 1 has to carry full weight of member.
4. Member may be lifted directly from the truck and rotated in the air.
5. Additional lifting loops or inserts are required at the top for stripping.

Fig. 2.2.10(f) Two-lines — sleeve at the top only.

2.3 Guying and Bracing

2.3.1 General

During construction, precast concrete members are exposed to temporary loads which include wind, seismic*, eccentric dead loads including construction loads, unbalanced conditions due to erection sequence, incomplete connections, and possible impact from construction equipment or impact from adjacent members while being erected. At this stage the designed structural framing is often not complete; therefore the installation of temporary shoring, bracing and/or connections are required to ensure stability of the structure. These measures should be addressed early in the building design stage. If such temporary techniques are used, the precast members should be provided with required inserts or weld plates for attachment of temporary works.

Erection stability can sometimes be built into the design. It may be possible to design the foundations and anchor bolts to withstand the forces generated from wind and temporary forces caused by construction procedures. **If base connections do not provide sufficient stability for columns and walls to be left free-standing, temporary guying and/or bracing shall be provided until final structural stability is achieved.** It is desirable to start erection from a laterally stable area such as a corner wall or stair tower.

2.3.2 Description of Methods

A wide range of bracing is available. For one- to three-story-high members, conventional steel pipe braces are most commonly used, Figs. 2.3.2(a) and (b). These devices combine strength, flexibility and adjustability into one unit. **Pipe braces shall be used as indicated in the bracing manufacturer's specifications regarding load, length and inclined angle. Special consideration shall be given to the location and the capacity of both the insert in the member and the deadman or ground anchor.** Anchors can be cast-in-place footings or slabs, drilled caissons, precast concrete weights or steel augers, see Fig. 2.3.2(c). Care should be taken to ensure that post-drilled anchors are installed at the correct embedment length. Temporary bracing should be arranged so as not to interfere with other members being erected. When long braces are used in compression, it may be necessary to provide lateral restraint to the brace to prevent buckling. Cable guys with turnbuckles are frequently required for higher structures; however, these are slow and expensive to install and are usually more cumbersome to work around on the site, Fig. 2.3.2(d). If used, long cable guys

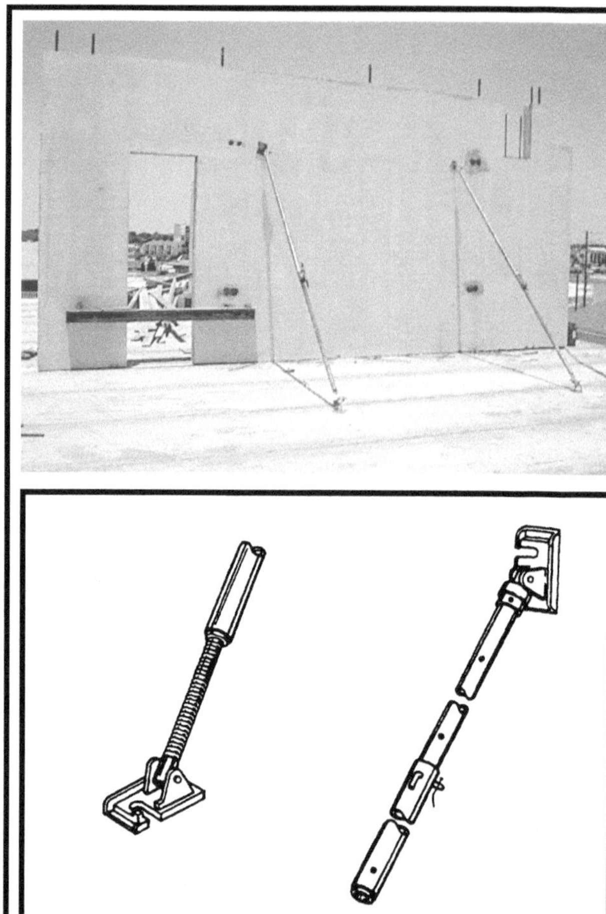

Fig. 2.3.2(a) Conventional steel pipe braces.

Fig. 2.3.2(b) Conventional steel pipe braces with lateral support to prevent buckling.

---

* Seismic forces are usually only considered when the condition of unsupported erection will be unduly prolonged (e.g., winter shutdown), or when dislodgments lead to total collapse or other catastrophic consequences.

Fig. 2.3.2(c) Precast concrete weights used as deadman.

Fig. 2.3.2(d) Cable guys used for lateral support of structure.

can stretch under load and should be adjusted as required. The minimum size wire rope used for bracing should be determined by a competent person and the turnbuckle should match the strength capacity of the rope. The wire rope can only resist tension, therefore, it is generally used in combination with other cable guys in an opposite direction or with compression struts which are often the precast components in the structure.

Consideration of the erection process should occur during the design of a precast concrete structure. Temporary connections to stabilize the partially completed structure should be considered as to their effect on the completed structure. The design documents should define the temporary connection details that are acceptable, and how they are to be incorporated in the completed structure.

Erection bracing may require the installation of temporary connections. It is not uncommon that these temporary connections are made part of the final connection, but they may behave in a different manner. For this reason, it is important that the designer be made aware of such intended usage, as well as the temporary forces induced, to permit proper evaluation. When connections are installed solely for erection, and restraint of rotation or movement can develop, they should either be removed once final stability is achieved or, if left in place, they should be disconnected. If this is not practical, the designer must be informed at inception, to permit proper evaluation. These connections should be noted on the erection drawings so that they are not part of the welding inspection process.

Removal of temporary bracing and shores shall not be permitted until proper alignment and adequate permanent support has been provided or until authorized to do so by the engineer of record.

The bracing must be maintained until permanent connections are accomplished. Single story load bearing panels should be temporarily braced until the connections are made and floor and roof structural elements are installed. Each precast concrete element should be braced independently of other elements, so that a given brace may be removed or adjusted without affecting adjacent members. Permanent connections which safely sustain loads imposed during erection may be used instead of temporary bracing.

When possible, the final connections should be used to provide at least part of the erection bracing, but additional bracing apparatus is frequently required to resist all of the temporary loads.

Diagonal cables, bracing columns, and connecting to shear walls and stair and elevator shafts, are the most common method of stabilizing the structure during the erection sequence. However, cables obstruct the continuing erection process and could become a hazard.

2.3.3 Responsibility

Responsibility for stability during erection shall be clearly established. Design for erection conditions must be in accordance with all local, state and federal regulations. Generally, procedures to ensure stability during erection such as bracing or temporary connection details should be developed by a professional engineer or competent person engaged by the precast concrete manufacturer or the erector. The design sometimes requires review by the engineer of record and/or building official.

Proper planning of the construction process is essential for efficient and safe erection. **Sequence of erection shall be established early, and the effects accounted for in the bracing analysis and the preparation of shop drawings.** Communication between the erector and precast concrete manufacturer is necessary in order to incorporate in the precast units all necessary attachments for guying and bracing, such as inserts or weld plates. See Section 1.2.

Erection drawings define the procedure on how to assemble the components into the final structure. When required, the erection drawings should also address the stability of the structure during construction and any temporary connections required. When temporary bracing, guying or shoring is required, additional bracing drawings are recommended. These should show items such as specific erection sequence, bracing hardware and procedures, and instructions on removal. **Shoring and bracing shall be installed as designed. Units shall never be left in an unsafe condition at any time.**

Attention should be given to possible deflections and/or rotation of structural members supporting precast concrete. This is particularly important for bearing on open web joists or cantilevered structural members. An example is a typical multi-story exterior column supporting spandrel beams loaded by long-span tees. Eccentric loads delivered to the columns by the beam reactions can cause column top deformations of 1 to 2 in. (25 to 50 mm) depending upon the structural components employed. Adequate temporary bracing may be required to prevent such deformations.

Structural steel beams, being relatively weak in torsion compared to concrete, generally require that the load be applied directly over the web or that additional bracing be provided to resist the effects of torsion. Problems during erection with rolling of steel beams may occur if they are not adequately braced.

Bracing analysis guidelines presented in Chapter 5 of the PCI Design Handbook, MNL-120.[4] For large and/or complex projects, a pre-job conference preceeding preparation of erection drawings may be warranted, in order to discuss erection methods and to coordinate with other trades. The crane selected may influence the erection sequence, and hence affect temporary bracing requirements.

Wind loads used for erection design should be based on local codes, tempered by engineering judgment.[5] For example, in hurricane regions, the maximum code load is usually not used, since there is sufficient warning of a hurricane so that additional bracing can be installed. Be aware that it is possible that more surface area is exposed to wind pressure during erection than when the structure is totally enclosed.

In seismic regions, the degree to which earthquake loads are considered for construction design is a decision which should be made by the individual designing the temporary bracing system, unless the subject is covered by local codes or project specifications. Often, seismic loading is neglected unless the uncompleted project is expected to be shut down in a temporary condition for an extended period of time.

Construction loads include materials stored on floor members such as masonry, drywall or other building materials, and construction equipment such as concrete placement buggies. A value of 25 psf (1.2 kPa) has been used without creating an excessive burden on the bracing requirements.

Fig. 2.3.3 shows other loading conditions which affect stability and bracing design.

2.4 Field Considerations for Connections

2.4.1 General Considerations

A major advantage of precast concrete construction is rapid erection of the structure. To fully realize this benefit and keep costs within reasonable limits, field connections should be kept as simple as possible, repetitious, consistent with adequate performance and ease of installation. To ensure the efficient erection of any precast concrete project, the type of connection details selected by the designer is a prime consideration. If there is any level of concern as to how the erection process can affect the connections, the designer should consult with the erector so that the sequences and procedures can be shown on the erection drawings. If the deflection of the structural frame is sensitive to the location or eccentricity of the connection, limits and/or adjustment mechanisms should be provided. Also, the precast concrete manufacturer or erector may prefer and be experienced with certain details and/or procedures not anticipated by the designer.

The erector should understand the function and performance of each connection detail to ensure that the precast units are installed in keeping with the design concept. **The erector shall be aware that the safe capacity of the**

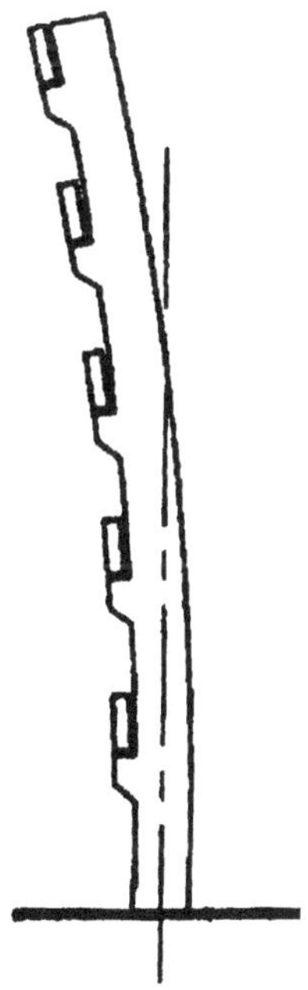

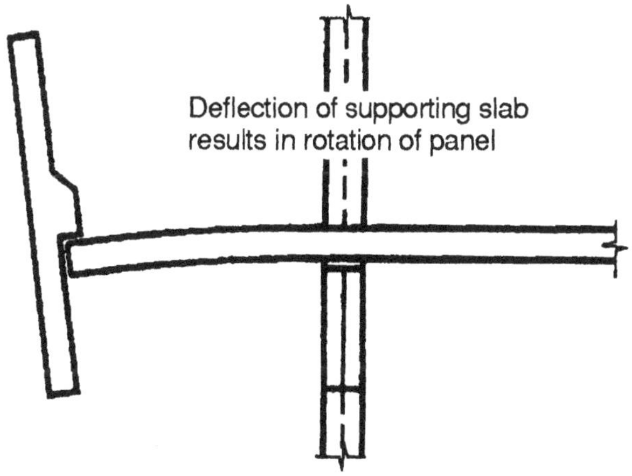

a. Columns with eccentric loads from other framing members will tend to lean out of plumb. A similar condition can exist when cladding panels are erected on one side of a multistory structure. Cable or other types of bracing can be used to keep the columns in position.

b. A classic example of rotation and deflection of framing members can be seen here where a panel is bearing on the end of a cantilevered floor member. Not only does the panel elevation drop, it also rotates out of plumb.

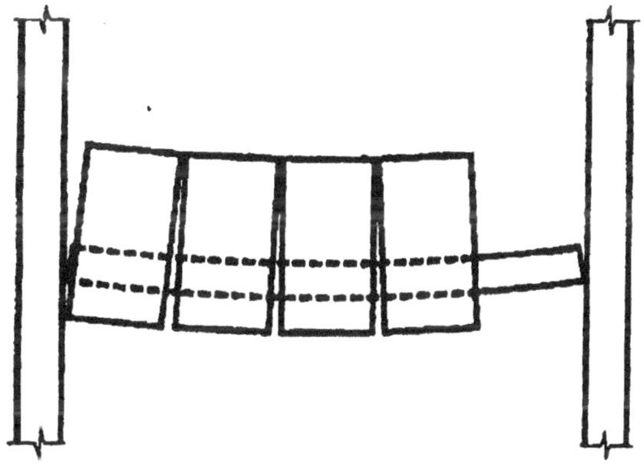

c. Deflection of the support member when loaded with the weight of the panels is another example of a situation which will affect the alignment of the panels.

The problems illustrated in 'a, b and c' are magnified with each additional piece set. There also could be loads, such as construction material or concrete topping, imposed beyond the weight of the precast that could further complicate the situation after all the precast is set.

The precast engineer should be able to help the erector determine how much deflection a given load condition will create. This way the erector can set the pieces higher so that when deflection occurs, the pieces actually will move into, or close to the proper position.

Fig. 2.3.3 Loading conditions which affect stability and bracing design.

designed connections may decrease with failure to comply with dimensional tolerance, incorrect use of hardware or bolt size, inadequate thread engagement of bolt, excessive weld length or improper bolt torque. Also, the combined shortening effects of creep, shrinkage and temperature change can cause severe stresses on precast concrete members and their supports especially if end connections restrain movement. **Each precast concrete unit shall be securely fastened in place as indicated on the erection drawings. Field checks shall be performed to ensure that connections have been installed in accordance with the approved erection drawings.**

Where necessary, the erector shall correct misalignment of the units resulting from deflection and/or rotation of the supporting structure due only to the weight of the precast concrete units. Units may be erected with a predetermined allowance for such movements or, following field installation, adjustments may be necessary to connections which are specifically designed for such movement.

A certain amount of field adjustment at the connections is normal. Product tolerances and site conditions preclude the possibility of a perfect fit in the field. This is true when the precast concrete pieces join to each other, and even more so, when the precast units interface with other materials. Each industry has its own recommended erection tolerances which apply when its products are used exclusively. The compatibility of those tolerances with the precast tolerances should be checked as part of the design process and adjusted when necessary (see p. 90).

Field adjustments or changes in connections which could result in additional stresses in the building structure, in precast products or in connections shall not be permitted without approval of the precast engineer. Particular care shall be taken to prevent damage to the precast concrete unit when adjustments are being made to move the unit into final position.

Operations which require working under a deck in an overhead position should be avoided. Therefore, the connections should be designed to permit operations to take place on the top side of erected members rather than from below where ladders or scaffolds are required, especially for welding. Availability and cost of equipment necessary to perform various operations such as welding, post-tensioning, or pressure grouting should be considered. Alternatives to any erection techniques which can eliminate temporary scaffolding should be considered. Access for wrenches on nuts turning through a large arc should be provided for bolted connections. When used, it is preferable that foundation piers extend above grade so that the anchor bolts can be adjusted during erection and the base plate need not be grouted in an excavation, which can fill with water and debris. Proper drypacking of column or wall panel bases in a narrow excavation is difficult.

All connections which serve similar functions within the building should be standardized as much as possible. As workers become familiar with the procedures required to make the connection, productivity is enhanced and the potential for error is reduced. Furthermore, standardization of details facilitates selection and shipment of connection items to the project with fewer delays and added economies.

Standardization also applies to dimensioning of connection details. Little is gained by slight changes in dimensions, since the savings in materials may be more than offset by the extra labor and possible errors involved in developing the modifications. If the changes in connection dimensions and materials are not in increments large enough for visual recognition, there is greater chance that an improper connection may be used at a given location.

It is desirable to have connections which are designed to enable the erector to safely secure the member to the structure in a minimal amount of time without totally completing the connections. If necessary, temporary bracing or connections should be used with final adjustment and alignment capability in all directions relative to the structure and adjacent components. The connection can be completed later independent of crane support. This allows the hoisting device to begin placing the next unit while connections on the first are being completed. The temporary connections should not interfere with or delay the placement of subsequent members. The temporary connections may have to be relieved or cut loose prior to completion of the permanent connections.

A potential problem in precast concrete structures is caused by the failure of the designer or erector to consider stability and equilibrium of the structure and its components, not only in its completed state, but during all phases of construction. An example is the L- or inverted tee-shaped ledger beam, Fig. 2.4.1(a). Because of eccentric loading, the beam is subject to torsion and tends to roll or rotate on its supports. The precast engineer should design connections for eccentrically loaded members that restrain the member from rotating. **If the erector discovers a situation in which rotational stability is not provided, the erector shall request the precast engineer to design a more stable connection.**

In some structures, restraint against torsional rotation in the completed structure is provided only by cast-in-place concrete which is placed after the precast members are in place. In these cases, temporary torsional restraint shall be provided during erection. This is frequently very difficult and costly to do. It is better if connections are designed to make the members torsionally stable during all phases of construction, as well as in the final condition.

Lateral stability is normally provided by shear walls, column-beam frame action or cross-bracing. Lateral forces

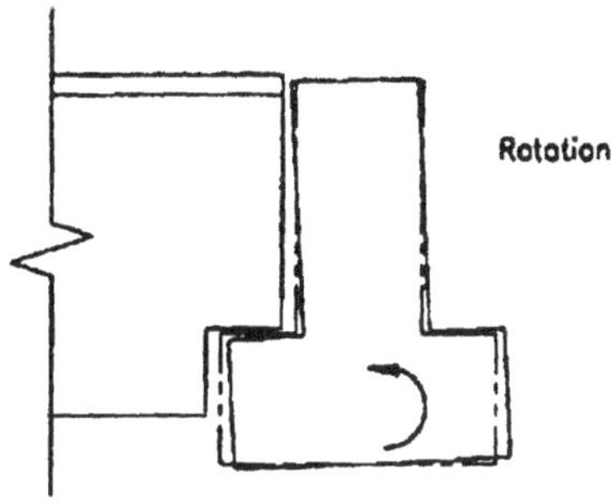

a. Unbalanced loading results in beam rotation.

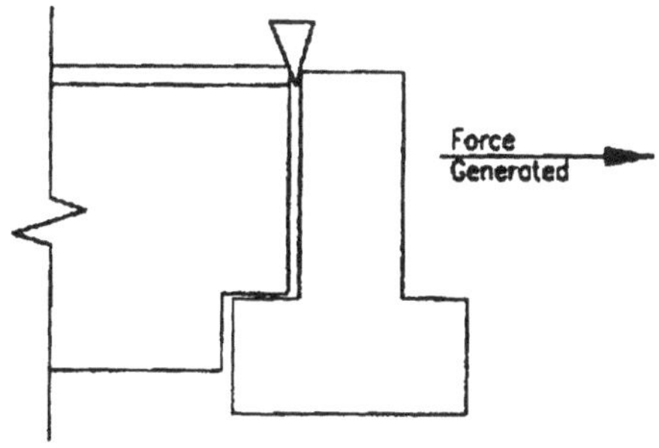

b. Driving a wood wedge between beam and tee flange. This method is not to be used in situations where tees or beam are on low friction pads. It is further recommended that this be done in conjunction with methods such as the ones shown in 'c and d.'

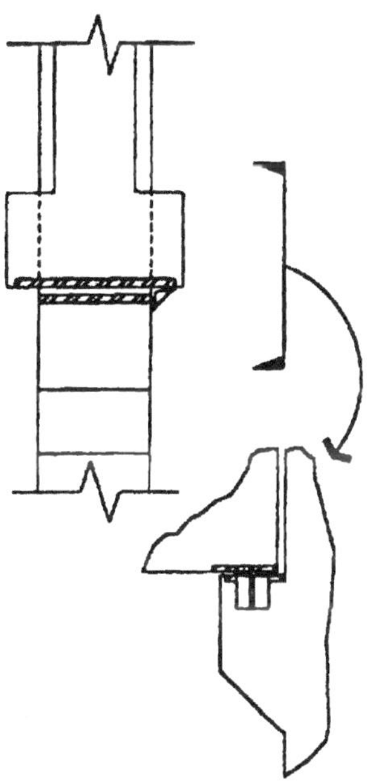

c. Welding an angle to the bearing plate of the beam and placed against the side of the support member on the loaded side. This solution works well in that the exact erection sequence does not need to be defined ahead of time.

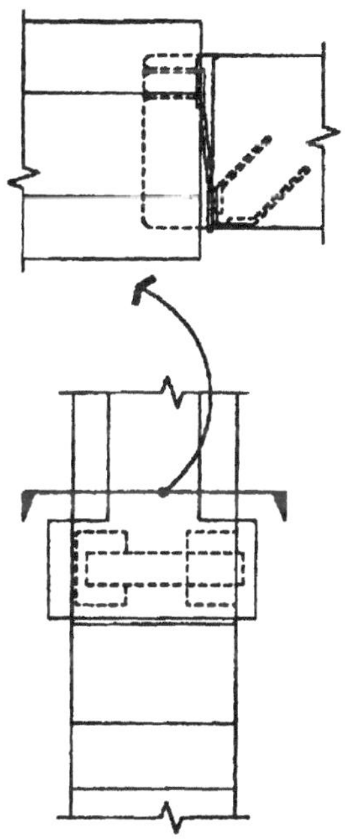

d. Pre-welding strap to the beam and then welding the strap to the embed in the support member. This solution works well in situations where the rotation loads are large. To use this system the erection sequencing needs to be defined ahead of time, or plates need to be doubled up to allow for setting in either direction.

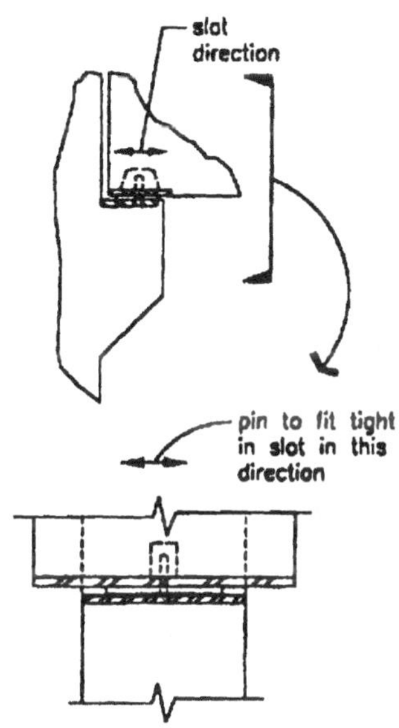

e. Bottom is restrained with pin projecting from corbel into slotted hole in bottom of beam. Top of beam is restrained by welding into column.

Fig. 2.4.1. Unbalanced loading due to partially complete erection may result in beam rotation as shown in detail 'a.' The most obvious solutions are to make sure that the beam is loaded evenly on both sides or to have the connections to the support member designed to take the rotation. This cannot always be done, in which case there are several options to minimize the rotation. One would be to shore under the loaded ledge to the floor below. Three other options are shown in details 'b, c, d and e.'

are distributed to the stabilizing assemblies through diaphragm action of floor and roof decks. **Temporary stability shall be provided and constantly maintained until diaphragms and connections designed to resist final lateral loads are in place.** In multi-story buildings this may require a detailed analysis and careful planning for all construction phases. **When connections are designed to provide some temporary moment resistance during erection, they shall be "released" (by removing bolts, or cutting welds) when the permanent lateral stability elements are in place.** If the erector does not release these mechanisms at the appropriate time, undesirable stresses could result.

Determination shall be made as to the number of floors that the erector may erect in a structure using temporary connections in order to make appropriate allowances for temporary stability. This is especially important for grouted connections, composite floor diaphragms, and cold weather conditions. For pre-topped structures, it may be necessary to use temporary shear ties from the pre-topped double tees to shear walls to produce erection stability if the final connections are achieved through closure pours.

The design engineer should be made aware of the cost and schedule consequences of connection details that do not allow erection to proceed irrespective of ambient temperatures without temporary protective measures. Materials such as grout, drypack, cast-in-place concrete, and epoxies may need protection or other special provisions when placed in cold weather. Also, welding is slower when the ambient temperature is low. If the connections are designed so that these processes must be completed before erection can continue, the cost of erection is increased and delays may result.

The supply of erection hardware, i.e., the loose hardware needed in the field for final connection of precast members, is normally the responsibility of the precast concrete manufacturer.

The responsibility for supply of contractor's hardware to be placed on or in the structure in order to receive the precast concrete units depends on the type of structure and varies with local practice, but in any case should be clearly defined in the contract documents. Hardware should be incorporated in the structure within the specified tolerances according to a predetermined and agreed upon schedule to avoid delays or interference with the precast erection. Hardware should be maintained and suitably protected (such as capping of inserts and protection of threads) until used by the precast erector. At times, it may be necessary to modify connections in the field to accommodate unforeseen clearance problems or the work of other trades. The GC/CM should discuss any modifications with the precast concrete manufacturer and the designer to ensure proper performance of the connections.

Care shall be taken to prevent chemicals, such as muriatic acid, from contacting the hardware and causing corrosion.

2.4.2 Types Of Connections

The designer should refer to the *PCI Design Handbook (MNL-120)*[4] and *Design and Typical Details of Connections for Precast and Prestressed Concrete (MNL-123)*[6,7] for connection design. Particular attention should be given to field adjustability. Hardware design for connections should take into account the tolerances for both the precast concrete components and the structure. These considerations may require clip angles and plates with slots (Fig. 2.4.2), or oversize holes to compensate for dimensional variations, field welding, or sufficient shim spaces to allow for elevation variations. Sufficient minimum clearance between precast units and structure should be provided to allow for product, interface, and erection tolerances. Hardware should be designed to compensate for additional stress at maximum anticipated clearance. Suggested clearances are given in Section 5.4.

2.4.2.1 Bolted Connections. Bolted connections simplify and speed-up the erection operation, because the connection is positive immediately. Final alignment and adjustment can be made later without tying up crane time. **Bolting shall be in accordance with the erection drawings, using material specified by the designer. A periodic check shall be made to verify that all bolts are of proper strength, length, thread and diameter.** For bolted connections, 1/2 in., 3/4 in., 7/8 in. or 1 in. (13mm, 19 mm, 22 mm or 25 mm) diameter bolts with National Coarse or coil threads are considered standard in the precast concrete industry and should be used. Regardless of the load requirements, a 1/2 in. (13 mm) diameter bolt should be the minimum used for any precast concrete connection. Coil thread stock or coil bolts may be used in lieu of National Coarse threaded stock to minimize the time required to make some connections and reduce the risk of thread damage.

The depth of the threaded recess in the unit shall be checked to assure minimum required thread engage-

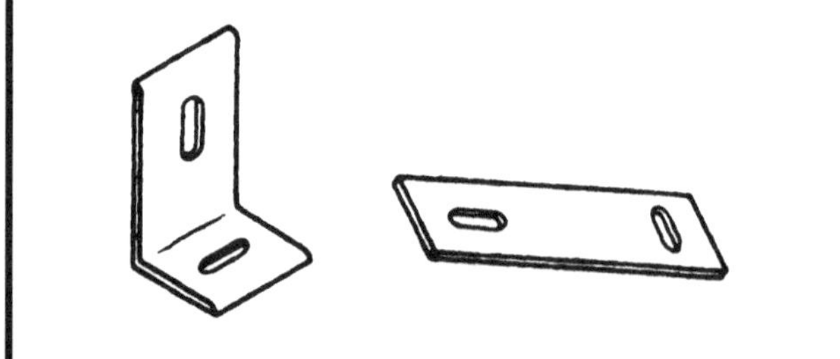

Fig. 2.4.2 Slots in clip angle and plate allow lateral (horizontal) and vertical movement. One end of clip angle or plate may be welded rather than bolted.

ment for the bolt. If the hole is too deep, there is a possibility that the threads of the bolt will not reach the threads of the insert cast into the unit. **A bolt that is too long shall be shimmed until it can be tightened without bottoming, and one too short shall be replaced with a bolt of proper length.** Care should be taken to prevent dirt, debris or ice from accumulating in the insert which may not allow full thread engagement.

Following erection of a precast unit where slotted connections are used, a check of bolt position and tightness shall be made. For sliding connections, the bolt shall be properly secured, but not tightened so that it can move within the connection slot. Steel washers when used to assure that tension has been developed in the bolt, shall be large enough to overlap the sides of the slots and allow for full movement. Such connections shall be coated to prevent rusting. Low-friction washers (teflon or nylon) may be necessary to ensure movement if rough surfaces occur between the elements of the connection. **When a friction connection has been designed to provide restraint, specified bolt torque shall be applied by the erector.**

Bolted connections should allow for industry erection tolerances. Clip angles should have slotted or oversized holes to allow for tolerances, adequate adjustment and movement due to expansion and contraction of the precast units and supporting members.

When the connection cannot be made because the insert is out of place or missing, alternative connections shall be developed and/or approved by the precast engineer. Expansion bolts may be used as corrective measures when cast-in inserts are misplaced or omitted. They are inserted into pre-drilled or self-drilled bored holes in hardened concrete. Strength is obtained through expanding parts of the insert which exert lateral pressure (wedge action) on the concrete. Lateral pressure is produced by tightening or torquing a bolt into the insert. Anchorage strength depends entirely on the lateral force. For connection reliability, the importance of correct installation and quality control cannot be overemphasized. Holes for expansion bolts should be drilled straight, of proper depth and diamenter, and cleaned. The bolts should be tightened. **The minimum distance to the edge of the concrete and spacing between bolts shall be based on the anchor manufacturer's recommendations.**

Expansion anchors when subjected to stress reversals, vibrations, or earthquake loading shall be approved by the engineer. Chemical anchors (resin capsule or epoxy anchors) could be considered for this application or for heavy loads. However, chemical anchors may degrade due to creep at temperatures in the range of 140 to 150 deg. F (60 to 66 deg. C). Such temperatures are readily experienced in warm climates, particularly in facade panels with dark aggregates. Also, chemical anchors may not be allowed in fire-rated connection assemblies. **Manufacturers' recommendations and local code provisions for installation of chemical anchors shall be followed.**

2.4.2.2 Welded Connections. Welded connections are the most common and typical connection used in the erection of precast concrete. They are structurally efficient and adjust easily to varying field conditions. The connections are usually made by placing a loose plate between two structural steel plates that are embedded in either the cast-in-place or the precast concrete, and welded together. Some connections are designed to bend and yield in one direction and be rigid in another while others are designed to be rigid in all directions. **Welded connections shall be installed exactly as shown on the erection drawings and details.**

Welded connections should be clearly detailed to show the type, size, length and location of all welds. **If this information is not available, the erector shall obtain the necessary information.** The erection drawing should identify non-critical erection welds that are for the sole purpose of stability during erection and are not required after the topping or other structural elements are in place.

The code that is most frequently specified for the welding done in the course of erecting precast concrete is ANSI/AWS D1.1.[8] This code also references specific information regarding electrode storage, weld joint preparation and preheating requirements.

There are, in addition to provisions of ANSI/AWS D1.1, precautions specific to the erection of precast concrete. An embedded plate should not be permitted to get hot enough that it cracks the concrete. A plate that is embedded in concrete will absorb heat at a slower rate than a loose plate being welded to the embedded plate. **The erector shall retain a weld testing agency or a welding consultant to establish their welding procedures and the certification test for their personnel.**

A welding procedure specification (WPS) shall be established for the certification of welding personnel. A WPS shows the type of joints, base metal, electrode size and type, position, heat treatment and preheating requirements, polarity, process, current, voltage, welding travel speed, technique, date, and the person authorizing the procedure. A sample welding procedure specification is shown in Appendix D. **If the welding required for a project differs from the conditions in the WPS, a new WPS shall be created and the welding personnel re-certified.** For galvanized connections, if the galvanized coating is removed so that the base material is the same as in the existing WPS, then a new WPS and welder re-certification is not required.

Once the WPS is established for the welded connections to be performed, a company shall certify their welding personnel to those procedures unless personnel are already certified for procedures that equal or exceed the company WPS. The certification states that the certified welder has passed a welding test and can perform a weld in conformance to the welding procedures specified. This certification is a welding performance qualification. The qualification test is not intended to be used as a guide for welding during actual construction, but rather to assess whether an individual has the required minimum level of skill. A welder can be certified in several prequalified welding procedures. ANSI/AWS D1.1 should be referred to for more specific information.

If no other means of inspection are provided, all welds shall be checked for completion by the erector. Welds that are critical to the structure of the building shall be visually checked by the erector. There are a number of defects that can be observed by a visual check such as porosity, undercut, too much or too little convexity, undersize, underlength or overlength. Verification of welds should include use of the proper electrode, proper storage of the electrode, and operator certification.

Care should be taken when welding in the vicinity of neoprene bearing pads to avoid damaging the pads. During welding the erector should protect adjacent concrete surfaces from damage resulting from weld splatter and smoke stains. Duct tape or soap on adjacent concrete can help protect those surfaces from stains. **When welding over installed or stored equipment or construction materials such as glass or aluminum, care shall be taken to protect materials from weld spatter.**

Gasoline, diesel or electric welders are all commonly used. **Care shall be taken when fueling the equipment when others are welding nearby. A fire extinguisher of proper size and type and in good operating condition shall be located within reach of all burning and welding operations.**

Refer to Appendix D — Welding, for more information on welding procedures.

2.4.2.3 Post-Tensioned Connections. Post-tensioning both vertically and horizontally may be used for field connections of precast members utilizing either bonded or unbonded tendons, Fig. 2.4.2.3.[9] Bonded tendons are installed in preformed voids or ducts and are made monolithic with the member and protected from corrosion by grouting after the tensioning operation is completed. Unbonded tendons are protected against corrosion by a properly applied corrosion preventive organic coating. Unbonded tendons are connected to the member only through the anchorage hardware, which also must be protected from corrosion.

Fig. 2.4.2.3 Ductile bars being tensioned using torque wrench.

Tendons consist of one or more wires, strands or bars made of high strength, high carbon prestressing steel susceptible to strength loss due to mechanical damage.[10] **Tendons shall be stored and handled to prevent excessive rusting or pitting, to avoid nicks or kinks, and to prevent damage from high temperature operations such as welding or flame cutting carried out near the tendons.** Bars should be stored straight, kept straight during handling, suitably supported to prevent excessive bending stresses, with any threaded portion adequately protected.

Short tendons should be anchored using anchorage devices on either rolled or cut threads in order to develop the required force. Wedge anchorages may be impractical because the anchor set may equal or exceed initial tendon elongation.

Anchorage devices for all post-tensioning systems shall be aligned with the axis of the tendon at the point of attachment; concrete surfaces, against which the anchorage devices bear, shall be normal to this line of direction or, if not, suitable wedge or bearing washers shall be used. Accurate measurements of anchorage losses due to seating and other causes, should be made and compared with the assumed losses shown in the post-tensioning schedule. When necessary, adjustments or corrections should be made in the operation.

Tendons may be installed in partial lengths and coupled together at special openings or at floor levels. Couplings should be used only at locations specifically indicated on the erection drawings or as approved. **If a tendon which**

contains couplers is to be stressed over the full length, the coupler shall be free to move. Shim packs supporting precast walls should be placed at post-tensioning areas only. If placed at other points the wall may crack if tensioning is done prior to drypacking or before drypack has achieved the required strength.

All shoring or bracing shall remain in place until the tensioning operation has been completed.

The instructions given on the erection drawings regarding post-tensioned connections should include:

1. The post-tensioning system to be used.
2. The maximum pressure to which the jacks, pumps, etc., may be exposed and/or the maximum force to be applied.
3. The area of the jack piston as assumed, if necessary.
4. Calibration of equipment.
5. The schedule and sequence of tensioning tendons as well as the required concrete strength of the precast units at tensioning.
6. Details, for each tendon, on how the tensioning should proceed. These details will usually include:
   a. Whether single or double-end tensioning should be used.
   b. The theoretical force or pressure which must be supplied by the jack, or the theoretical pressure gauge reading to be obtained.
   c. The corresponding calculated elongation of the tendon.
   d. If desired, the steps by which the pressure is to be increased during tensioning, and the maximum allowable pressure.
   e. The limits within which the results of the tensioning operations can be considered satisfactory, and precautions to be taken in case the results are not within these limits.
   f. Recording of results and procedures for final approval of the tensioning.

Post-tensioning operations call for extreme care. **Personnel involved in the tensioning operation shall be properly qualified and experienced with the tensioning procedures and the equipment to be used. Concrete and joint grout or drypack strength at time of post-tensioning shall exceed the minimum required by the precast engineer prior to tensioning.**

No one shall be permitted to stand directly behind the jack or around the dead end of the member being tensioned. All personnel shall be aware that tensioning is about to begin. Audible and visible signals shall be employed and remain during tensioning. Any personnel not directly involved in the tensioning operations shall be cleared from the designated area and remain clear until audible and visible signals cease.

Grout may be required to provide corrosion protection to the tendons and/or to develop bond between the tendons and the surrounding concrete. **Grout shall fill all the voids in and around the tendon for its entire length. Grouting procedures shall follow the "Recommended Practice for Grouting of Post-Tensioned Prestressed Concrete."** [11]

If weather conditions or scheduling of job require grouting before tensioning; (1) the post-tensioning tendon shall be greased and wrapped and free to elongate, and (2) a suitable gasket shall be placed on top of nuts and couplers (couplers to be wrapped and greased also) and be twice as long as the expected elongation or factory installed thread protector left on tendon. In certain cases, a chemical retarding agent may be added to the grout mix to assure that the tensioning will be performed while the grout is still in the liquid stage, and eliminate the need for greasing and wrapping the tendon.

If weather conditions or scheduling of job require grouting after tensioning; (1) tendons do not have to be greased and wrapped, and (2) vent tubes for conduits shall be in place at top and bottom of unit.

2.4.2.4 Dowel/Anchor Bolt Connections. In a dowel connection, the strength of dowels in tension or shear depends on dowel diameter, embedded length and the bond developed. Good practice is to provide sufficient embedment to develop the full dowel strength.

Threaded anchor bolts and rebar anchor dowels which protrude from the foundation are the critical first connection to precast members. Usually, this work is performed by a subcontractor not responsible to the erector. It is important that these items be placed accurately in both plan and vertical alignment.

The precaster should furnish the foundation contractor with a template for locating the bolt or dowel pattern. This template is used as a master for the jigging system, which can be shop fabricated or proprietary. The key is to make a jig which holds the bolts or dowels firmly in plan and vertical alignment.

The jigs are surveyed twice. The first time is prior to casting the foundation. Then, just after the pour, the jigs should be surveyed again to check for displacement due to form movement.

Reinforcing bar anchor dowels can be cast-in overlength 4 in. (100 mm) or more. The extra length can be cut off later as final elevations are set and shims are placed. Anchor bolt projections need to be the correct length.

There are two types of dowel connections: generic and proprietary. In some cases, both types may be used on the same project, even near each other. It is important that the mixture and placement of the grout be carefully controlled. Quality control inspectors should be on hand to monitor the operation and sample the grout.

Dowel connections are not structurally sound until the grout has cured, usually overnight for required minimum strength and 7 to 28 days for full design strength. **A placement or pumping procedure shall be employed that ensures that grout completely fills the tube or sleeve.**

Pre-blended and packaged grout mixes shall be used. The proprietary splice sleeve systems require that specific grouts be used. **The erector shall not substitute grouts without written authorization. When using proprietary grouts, preparation of the sleeve, mixing of the grout and installation shall be done according to the manufacturer's instructions.**

Cold weather can be a factor. Most grouts are not designed to be placed below 40 deg. F (4 deg. C) with 50 deg. F (10 deg. C) as the practical minimum temperature. The colder the grout, the slower it cures. It may take several days in the winter. If freezing weather is anticipated, external heat should be applied to the grout after units are erected.

If the temperature of the precast member is below freezing, it shall be warmed before grouting. This can be done by tenting the area and aiming propane or electric radiant heaters at the surface near the connection. Electric blankets are also effective where tenting is impractical.

The best practice is to raise the temperature of the interior of the tube or splice sleeve to at least 60 deg. F (16 deg. C) before grouting, and to keep the heat applied overnight or until cube tests show that minimum strength has been reached. Being too warm is usually not a problem. Heat serves to accelerate the strength gain of the grout.

Cutting torches shall not be used to heat the tube or sleeve area. Hot spots will cause entrapped water in the concrete to boil and cause the concrete to spall. With radiant heaters, it usually takes only an hour or two to bring the temperature in the tube or sleeve to the required minimum.

In hot weather, it is not necessary to cool the tube or sleeve, but the grout mix shall be kept below 80 deg. F (27 deg. C) to guard against flash setting in the pump. The grout packages should be stored in the shade. The packages needed for the following day's work should be placed in an air conditioned trailer overnight. Cold water from an insulated cooler should be used for the mix. Normally, ice should not be added directly into the mix.

Erection drawings should clearly show the strength to be obtained in the grout before erection may proceed and at 28 days. Follow the instructions on the grout packages for achieving the specified strength. Equally as important is the strength to be obtained before bracing is removed. Two in. (50 mm) cube molds come in sets of three. Cubes should be taken each day and tested at 24 hours (or whenever the connection is loaded) and at 28 days to verify the strength of the grout. Sampling techniques are different for grout cubes, and some proprietary grouts have special procedures and molds. Be certain that technicians are using the correct methods and equipment.

Reinforcing bar dowels should be straight and embedded the proper length. Lengths should be clearly shown on the erection drawings. Every precaution should be taken to prevent the dowels from being bent during shipping and erection. Should a dowel require straightening, follow the procedures recommended by the American Concrete Institute (ACI 301-96, Section 3.3.2.8)[12] to minimize the effects of cold working and fatiguing the metal.

Dowel holes or sleeves should have reasonable clearances for rapid erection. The edges of the dowels should be beveled when they are to pass through corrugated tubing. Otherwise, the sharp edges will catch on the corrugations making it difficult to fit the pieces together quickly.

When dowels are projecting down, it is likely the tube or sleeve will need to be filled just prior to the piece being set. This is called a "pre-grout" situation. To assist in aligning dowels with sleeves, it is desirable to have the dowels longer in the outside corners or ends. In handling a precast piece with downward projecting dowels care should be taken to avoid inserting the dowels in dirt which can contaminate the connection and could cause the dowel to lose bond with the grout. Care should be taken to prevent the dowel holes from filling with debris or with water and freezing.

The erector shall coordinate the grouting in a pre-grout situation as the pieces are erected instead of having a separate operation following this work. Many erectors test-fit the piece and then lift it out and away so that grout can be poured into the sleeve. **The piece shall not be suspended over workers while they are placing grout.** Grout level and consistency should permit displacement of some grout when the unit is placed.

The advantage of this arrangement is that the connection is completely hidden with no exposed hardware or grout tubes to be patched. Winter conditions may require heating, which can slow the erection process significantly.

When the dowels are projecting up into a sleeve or tube, this is called a "post-grout" situation, Fig. 2.4.2.4(a). Grout must be pumped into the sleeve after the lower and upper precast pieces are set. There should be a lower and upper port to ensure that grout has flowed completely through the tube. In the winter, these connections can be heated and grouted after erection, so efficiency is not reduced significantly.

Grout is pumped into the lower port and should flow out of the upper port to ensure all air voids have been displaced, Fig. 2.4.2.4(b). Pump a few extra strokes and catch the overflow in a cup which can be returned to the hopper for reuse. After the flow subsides, but before removing the nozzle from the lower port, seal the upper port to create a closed system. Then, remove the nozzle and quickly seal the lower port.

Rubber stoppers are best for sealing ports. Many erectors use short lengths of closed-cell foam backer rod force-fit in the tubes. Do not use open-cell backer rod as it will act like a sponge and siphon mix water away from the grout causing lower strength.

If a dowel is installed in a horizontal hole, the hole shall be sealed and pressure injected.

When proprietary grout-filled splice sleeve systems are used, they shall be installed in accordance with the manufacturer's recommendations.

Fig. 2.4.2.4(a) Use of projecting dowels and post-grout tubes.

Fig. 2.4.2.4(b) Pumping of post-grout tubes.

When reinforcing bars are used in a corrugated tube, the sides of the tube should be free from form oil, grease, loose concrete and dirt in order to achieve maximum bond with the grout and the surrounding concrete. Just prior to grouting, the concrete surface inside the tube, if exposed, should be in a saturated, surface-dry condition.

Precast unit adjustment made after the initial set of the grout may destroy dowel bond and reduce connection strength. If protruding dowels are expected to cause difficulty in shipping or rotation of units on a trailer, then proprietary "dowel bar substitutes" can be used in the precast member with threaded "dowel-in's" screwed into the receptacle in the field just prior to setting. Be certain the product chosen fully develops the entire capacity of the reinforcing bar, because some threaded systems cut into the body of the bar and reduce its strength.

Units that are required to have grout between them can be either set on shim stacks and then grouted later or set immediately in fresh grout with joint width and elevation controlled by previously placed shim stacks. Joint grout may be placed or pre-bedded just prior to setting the precast piece. The joint grout should be mounded slightly and excess material trimmed after setting. Alternately, a form can be made and the joint pumped-full later. Dry-packing is the third method to complete the joint. It requires greater than a 1 inch (25 mm) joint dimension for best results.

The maximum number of units which can be set using shims only should be determined prior to start of erection. **Care shall be taken to ensure that the grout completely fills the joint void space to prevent high stress concentrations on shim packs.**

Joint grout should be kept out of the sleeves in proprietary systems. The grout in these systems has special high-strength properties and sometimes contains ferrous aggregate.

2.4.2.5 Grout, Mortar And Drypack. Grouts and mortars are used primarily for load transfer and to patch or fill voids. When patches or non-structural void filling is the objective, job-mixed grouts or mortars are acceptable. **When grouts are used to achieve positive, non-shrink, load transfer properties, proprietary non-shrink or factory-controlled products shall be used or sufficient testing performed on job-mixed grouts to assure uniform results. Care shall be taken not to use ferrous expanding agents when joints are exposed.**

Drypack is a term describing a method of placing rather than a type of material. Drypacking uses only enough water in the mix to produce a stiff, but damp, granular material which must be packed into place. Both proprietary and job-mixed grouts or mortars can be used for drypacking. However, the proprietary grouts and mortars will generally yield higher compressive strengths. This is due in part to the selection and quality control of raw materials and the inclusion of special additives. If aggregates are added to the grout or drypack, they must be clean, dry, graded and of rounded or spherical shape.

Joints that require drypack or non-shrink grout for final completion should have a minimum thickness of one inch as the planned dimension between two surfaces, 2-1/2 in. (63 mm) is preferable for grouted base plates under precast columns.

A grout pump is most efficient for grout placements of hollow-core slab keyways requiring from two to twenty cubic yards and up to twelve stories height. When used on lifts of over four stories, it is recommended that the vertical section of hose be replaced with steel piping strapped to the side of the building to reduce as much hose friction as possible. On projects requiring less than two cubic yards of grout, it is generally more efficient to mix the grout in a portable mixer, move it to the deck in a wheel barrow either directly or by crane, and bucket or pour it into the keyway joints of hollow-core slabs. Projects at greater heights (over 10 stories) are best grouted with a crane equipped with a 1/3 to 1/2 cubic yard (0.25 to 0.4 m^3) concrete bucket, Georgia Buggies, or ordinary buckets for pouring the grout in the keyway.

In general, grouting and drypacking should proceed as members are erected and composite topping should not be more than three levels behind erection in any one area. Minimum topping pour sizes must be considered, however. Composite topping typically is placed on top levels first, proceeding to lower floors, in multi-bay pours. Topping usually takes place after the deck has been erected completely or to a natural dividing point such as an expansion joint.

The erector shall consider the following items when grouting:

Fig. 2.4.2.5(a) Leveling of hollow-core slabs prior to grouting.

1. **All concrete elements shall be cleaned of coatings and miscellaneous debris. They shall be thoroughly wetted in order to prevent water from being absorbed out of the grout mix.** Metal surfaces should also be wet before grouting to facilitate the flow of grout around and under bolts and plates.

2. **All precast members shall be aligned and in their final position prior to grouting, Fig. 2.4.2.5(a).**

3. **Forming for joints with flowable grout shall be installed to minimize leakage and resist grout pumping pressure.** With a pump, grout is forced into a hollow-core slab keyway allowing the grout to rise 1/8 to 1/4 in. (3 to 6 mm) above the plank surface to compensate for grout shrinkage.

4. **The initial set time of the grout shall be determined in accordance with ASTM C191[13] and be appropriate for the placing operations recognizing the ambient temperatures at the time of grouting.**

5. **Precast members with dowel connections shall be aligned in the dowel pockets or sleeves prior to initial set of the grout.**

6. Precast members required to be set in a bed of grout can either be set on shim stacks or leveling nuts, and then grouted later, or they may be set immediately in fresh grout with the elevation controlled by previously placed shim stacks or leveling nuts, Fig. 2.4.2.5(b). **The maximum number of load-bearing wall panels that can be set using shims only shall be determined from the erection drawings or engineer prior to start of erection.** Care should be taken to ensure that

Fig. 2.4.2.5(b) Leveling nuts and shim stack ready for erection.

the grout completely fills the joint void space to prevent high stress concentrations on shim packs. The grout joint should be flush or tooled so as not to interfere with any baseboard or wall covering.

7. **Flat slabs and hollow-core slabs shall be leveled, large shear key openings blocked, and all shear key reinforcing bars placed prior to grouting.** In leveling slabs, care should be taken to keep the bottom of the slabs level. It is much easier to "feather" the top surface than it is to fit room partitions around an uneven ceiling. After grouting the shear keys, excess grout should be struck from the bottoms of the joints and, if required by contract, the top "feathered" as required for the specified floor covering. Grouting should be a continuous operation and each level of slabs should be grouted completely prior to continuation of erection, Fig. 2.4.2.5(c).

Fig. 2.4.2.5(c) Grouting of shear keys and butt joints in hollow-core slabs.

Fig. 2.4.2.5(d) Base of panel is grouted after all adjustments are made.

8. Column base plates, in most cases, should be grouted or adequately shimmed prior to loading the frame with precast decking, Fig. 2.4.2.5(d).

9. Cold weather grouting is critical due to the possibility of grout freezing and its adverse effect on the structural properties of the installation. Grouting is generally permitted when the ambient temperature is 40 deg. F (4 deg. C) and rising. Concrete or grout strength increases very slowly if exposed to low temperatures. Cold weather protection and additional curing are necessary if ambient temperatures are below 40 deg. F (4 deg. C), Fig. 2.4.2.5(e). A space enclosure and heating should be used if early strength is required. The use of additives to increase early set and strength is advantageous, but must be carefully controlled. Grouts should not be placed against concrete with a temperature at or below freezing.

10. **Calcium chloride shall not be used as an additive in grout when grout comes into contact with metal parts or reinforcing steel in a connection.** A proprietary, gypsum-based portland cement may be used to

Fig. 2.4.2.5(e) Cold weather protection of grout at base of column.

provide high-early age strengths under cold weather conditions. However, this material must be used only on above-grade applications since prolonged exposure to water will cause strength loss (it is partially water soluble) and long-term expansion. **The proprietary, gypsum-based portland cement and other special materials, such as expansive cement, high early strength grout, epoxy grout, etc., shall always be used in accordance with manufacturer's recommendations and instructions.**

Conduits, pockets, blockouts, sleeves, bolt holes or inserts may be a water trap. A water trap can be a potential problem and should be protected or free immediate drainage should be provided especially for winter erection. **If free drainage is not possible, areas likely to trap water shall be protected immediately after erection using grout, plugs or covers of impermeable material.**

Grouting operations can trap water in conduits or bolt sleeves if drainage or vent holes are plugged. Also, if grout is not mixed thoroughly segregation of water and grout can occur. During the first winter, frozen water could burst these conduits and sleeves if panels are exposed to freezing temperatures. It is extremely important to mix grout thoroughly.

2.5 Installation

2.5.1 General

Erection procedures for the various types of precast and prestressed concrete members vary depending on the size, shape and design of the units, the structural elements which will receive or support them, and the overall complexity of the structure.

Prior to start of erection, final review shall be made to ensure that all materials and supporting equipment have been ordered and are either on the jobsite or readily available. This includes shims, bearing pads, expansion joint materials, connecting hardware, welding rods, grout, braces, guy lines, rigging, tools, and necessary equipment. **A final review shall be made for hoisting equipment radii, size and capacity and, if necessary, special permits and verificaton of current crane inspection.**

A review shall be made of the erection drawings, bracing drawings, written procedures and calculations related to shoring, bracing and guying to meet wind, eccentric loadings, dead, live and superimposed loads applied during erection.

Hoisting precast pieces is usually the most expensive and time-consuming process of erection. Connections should be designed so that each unit can be lifted, set, connected, and unhooked in the shortest possible time. **Before the unit is unhooked, it shall be stable and safely in position.** Some units, such as double tees and hollow-core slabs, are inherently stable and require no additional connections before releasing the crane. Others, such as columns, deep beams, wall panels and single tees usually require some supplemental shoring, guying or fastening before the crane can be unhooked. Preplanning for the fewest, quickest and safest possible operations that should be performed before releasing the crane will greatly increase efficiency of erection. Bearing pads, shims, or other devices upon which the piece is to be set should be placed ahead of hoisting. Loose hardware that is required for the connection should be readily available for attachment.

In some cases, it may be necessary to provide temporary fasteners or leveling devices, with permanent connection made after the crane is released. For example, if the permanent connection requires field welding, grouting, drypacking, or cast-in-place concrete, the use of erection bolts, C-clamps, guy lines, pins, or shims should be considered. **Temporary fasteners or leveling devices shall be given careful attention to ensure that they will secure the piece in its proper position during the placement of adjacent pieces that are erected before the final connection is made.**

Shims are often used as spacers or as means of leveling or aligning adjacent components. They serve as an interim or temporary load transfer device. **Unless temporary loading of units has been specifically incorporated or allowed in the contract documents, the erector shall be responsible for the temporary bearing and loading of units.** High-density plastic and steel shims are commonly used to attain the specified joint dimension. Shims should be placed away from the exposed face of the unit to prevent spalling in the event of excessive loading. Also, they should be recessed out of view as they can be unsightly and difficult to remove if left exposed to view. **Temporary shims shall be removed from joints of non-loadbearing units after connections are completed and before applying sealant unless the shim material is readily deformable.**

For loadbearing units, the use of non-metallic shims or bearing pads such as neoprene or plastic may be used where bearing pressures do not exceed the allowable concrete stresses and the potential risk of damage to concrete edges is negligible. If left in place, shims should be noncorrosive or protected so that staining will not occur.

The indiscriminate use of shims, particularly steel shims, can sometimes lead to undesirable consequences. For example, steel shims have been used in conjunction with grouted joints in multi-story bearing wall construction. Even with well-compacted grout or drypack, the compressive modulus of the steel shim is six times that of the drypack; consequently the grout will deform (compress) more readily than the shim. The principal load transfer path will remain concentrated through the steel shim rather than distributed along the grout bed. High load concentrations at shims can cause spalling at panel surfaces or crack panels vertically.

In other circumstances, the unplanned use of shims may interfere with the component interaction intended by the designer. They may prevent the independent behavior of the connection mechanisms and/or cause transfer of load where none was intended.

The erector shall consult with the precast engineer regarding the use of shims that are not specified on the erection drawings or have an excessive shim stack height.

Bearing pads are used to distribute loads over the bearing area and will compensate for slight bearing surface irregularities. Some types of pads reduce force build-up at the connection by permitting small displacements and rotations. Physical characteristics of bearing pad material necessary to satisfy these functions are: (1) permanent and stable; (2) able to equalize uneven surfaces and avoid point pressure; and (3) accommodate movements.

Erection drawings should indicate clearly the type, location and orientation of all bearing pads. The two types of materials most satisfactory are: (1) elastomerics with known compression, shear and friction strength and ability to deform with movements; and (2) plastics with low friction coefficients and high compression and shear strength. **Bearing pads shall be properly placed to transfer the bearing stresses into the supporting structure. The final position of the pads shall be closely monitored during erection.**

Precast concrete units shall be erected at the locations shown on the erection drawings within the allowable tolerances. They shall be positioned so that cumulative dimensional errors do not exceed allowable tolerances. Horizontal and vertical joints shall be correctly aligned and uniform joint width shall be maintained as erection progresses.

2.5.2 Loadbearing Members

2.5.2.1 General. Consideration for the stability of the structure under construction is required during planning for erection.

2.5.2.2 Columns. When using double nuts and plate washers, the lower nuts can be preset to elevation, Fig. 2.5.2.2(a). With the columns in place, the upper nuts are installed and the crane is unhooked. Leveling nuts provide the simplest method for erecting columns since the full load of the column can be let off the crane and the column plumbed by adjusting the level of the nuts. The column is

Fig. 2.5.2.2(a) Column base plate and shim stack.

braced and then the crane is released. The base plate and double nut method is high in hardware costs, but provides tolerance for out-of-place anchor bolts as the base plate holes may be slotted or made oversize.

The preferred procedure for heavy loads is a column base plate with a single shim pack centered under the column. The shims take full load and the anchor bolts and nuts serve to stabilize the column. An advantage is that the columns can be loaded immediately.

Anchor bolts which project from cast-in-place foundations should be protected from damage and rust.

Sufficient access is required to tighten anchor bolts in base plates. **If the column is to be loaded immediately, sufficient shims shall be installed to transfer the load.** Although some shims might still be necessary for elevation adjustment, if the columns are erected far enough ahead of the rest of the structure, drypacking or grouting of bases with non-shrink grout can follow immediately after column erection eliminating the need to rely on shims for load transfer, Fig. 2.5.2.2(b). Care should be taken to maintain the columns plumb. If anchor bolts are designed for column restraint, all nuts should be retightened after columns are loaded because shims or grout can compress and nuts might loosen. Continuous-thread, high-strength bars and nuts may be used instead of standard anchor bolts.

Fig. 2.5.2.2(b) Drypacked column base and bolt recesses.

If the design calls for a base detail other than using anchor bolts, the correct elevation is established at top of shim pack and the column is stabilized and held plumb with two braces placed at 90 deg. to each other, Fig. 2.5.2.2(c). Long columns placed in this manner may have to be guyed or require long, strong shores.

The precast concrete manufacturer's instructions shall be followed closely in the unloading and turning of columns. Columns up to 40 ft (12 m) in length can normally be picked up directly from the truck with one crane load line hooked to a lifting loop in the top of the column or to a pin and lifting frame through a hole in the column. Columns in excess of 40 ft (12 m) will generally require turning in the air with two lines, one attached at the top of the column and the second line connected through a rolling block attached at the fifth point of the column. Columns in the 60 to 90 ft (18 to 27 m) length range require further special handling for hauling and erection and a rolling block system is required for turning into the vertical position. Guying or heavy shoring is normally required for columns in excess of 60 ft (27 m). Long columns can be spliced when it is not practical to ship or erect in one piece.

Multi-story precast concrete columns usually have guy cables and/or pipe braces which allow the erector to plumb and stabilize the column prior to, during and after the installation of upper level beams and floor slabs. As beams and floor slabs are erected, columns frequently start to bow or lean due to eccentric loading. If this bowing or leaning is not kept in control, the upper level floor members will not be able to be fitted in place. To control inward displacement as the various levels are erected; (1) hardwood wedges may be driven between the ends of members at the supporting beam; (2) adjustments can be made with guy cables and/or pipe braces; and (3) permanent connections can be made.

Monitoring column plumbness is essential as each level is erected. Columns should be plumbed by using two transits placed 90 deg. to each other. **Column plumbness shall be rechecked after applying each subsequent eccentric load or if the column is bumped while erecting other units.** Periodic checking of the column plumbness should be made throughout the erection process. Checking of column plumbness and providing adequate shoring are particularly important when using pretopped tees. The heavier load of the pretopped members increases the tendency for columns to come out of plumb. To maintain a plumb condition, columns should be braced and shored.

Fig. 2.5.2.2(c) Braces at 90 deg. to each other to stabilize column.

If columns are eccentrically loaded, they are often erected slightly out of plumb on the side opposite the eccentric load so that the eccentric load will bring them back to plumb when fully loaded. The amount of "out-of-plumb" is determined by trial and error for each project and type of column. The column size, height and design affect its deflection under load.

2.5.2.3 Beams. Beams are generally erected on columns, column corbels or walls. If the columns and walls have been erected in their correct locations, further layout for the beams should not be required other than keeping joint widths constant. Beam to column connections will vary depending on design requirements and/or architectural details. A bearing material, usually a reinforced masticated rubber pad or neoprene, should be used to provide uniform bearing for the spandrel to column bearing condition. Care should be taken to avoid rigidly fixing beams to columns, such as welding both the top and bottom of the beam. Flexure of the beam could result in distress in the connection. A connection that does not require immediate grouting is desirable, as unfavorable weather could delay erection.

The most common method of lifting beams is by use of spreaders hooked to the top of the beam. Stability must always be considered during erection of beams. If the beam tips sidewise, it may become unstable and cause serious danger to workers and structure below. Since it is difficult to move beams once placed, they should, if possible, hang plumb and level from the crane hook and be set in proper location initially, Fig. 2.5.2.3(a). If a member hangs out of plumb, a levelling device should be used to rotate the beam into its required position.

Inverted T-beams and L-shaped ledger beams shall be checked for rotation when loaded eccentrically, see Fig. 2.4.1. Wedging between deck members and the vertical webs of the beams may help prevent rotation. **Horizontal restraint in the direction opposite to rotation shall be provided at the beam supports to prevent unintended beam displacements.** Care should be taken to recheck the columns for plumb. If only one side of an inverted T-beam is temporarily loaded, it may be necessary to install temporary shores under the loaded edge until balanced loading is accomplished, or the designer should be requested to suggest temporary connections. The load should be kept fairly uniform on both sides of the T-beam

Fig. 2.5.2.3(a) Typical beam erection.

by placing slabs alternately on opposite sides to prevent twisting or rotation.

Spreaders of proper length shall be selected for picking up long, slender beams in order to prevent buckling. If this danger exists, it is better to use two cranes with a vertical load line at each end, Fig. 2.5.2.3(b). Eccentrically applied loads can cause rotation and buckling. Temporary guying or bracing may be required for stability until slender beams are finally secured in the structure. If stiffening trusses are used to help place such beams, care should be taken in removing the trusses so they do not interfere with previously placed units.

2.5.2.4 Wall Panels. Loadbearing wall panels may be precast or prestressed concrete units, stemmed, solid or hollow-core members. These members may be sandwich insulated panels and may have an architectural finish. **Wall panels shall be handled with care in order to prevent soiling, cracking, chipping or spalling of exposed surfaces.**

Loadbearing wall panels should be plumbed using transits. While the crane is holding the panel, an adjustable push-pull brace should be installed, see Fig. 2.3.2.(a).

Fig. 2.5.2.3(b) Tandem lift of long beams with two cranes.

Inserts for attaching bracing are normally cast into the backs of the panels, but may be field-drilled.

Base connections for multi-story loadbearing panels are normally achieved by weld plates or grouted mechanical sleeves. The walls may also be post-tensioned vertically using high tensile strength bars that couple at pockets at the base of panels at each floor level. The base of the panels should be grouted or drypacked before the next level of floor slabs is erected, Fig. 2.5.2.4(a). **Panel bases shall be grouted or drypacked prior to vertical post-tensioning unless sufficient shims are provided.** The grout should achieve its required strength prior to post-tensioning. For panels which are supported on a continuous bed of grout the maximum number of floor levels which can be erected using only shims should be determined by the engineer and, if critical, indicated on the erection drawings.

Structural stability is normally achieved through floor and roof diaphragms with high-capacity connections at stair walls, elevator shafts or other shear wall locations. **All bracing shall remain in place until stability is achieved, Fig. 2.5.2.4(b).**

Supports for floor or roof slabs are normally achieved by casting corbels or recesses on the interior face of wall panels at the desired level. If stemmed members are also used for floor or roof spans, the stem spacing would normally be the same as for the stemmed walls, therefore supports could be individual corbels, steel knife edges or other suitable brackets cast into the stems of the vertical members. Special considerations should be given to long spans where connections are designed to allow for expansion and contraction movements and rotations.

Two-line turning of stemmed members may be required, depending on their length and design. Details showing proper procedure for two-line turning may be found on pg. 23. The bottom edge of the members may require

Fig. 2.5.2.4(a) Drypacking of base of panel.

Fig. 2.5.2.4(b) Walls temporarily braced.

Fig. 2.5.2.4(c) Wall panel hoisted from the top.

protection devices during rotation into the vertical position. Accurate spacing of stemmed members is required in order to achieve acceptable visual appearance of joints and to achieve correct overall building dimensions. Variation in camber and variation due to warpage may be adjusted-out by use of flange connections.

Hollow-core or solid wall panels delivered in the flat position may be lifted off the truck in a horizontal position and rotated into the vertical position using one or two lines. If one line is used a protective device at the base to prevent spalling during rotation may be required. Panels delivered on edge require two-line turning. A specially designed turning device needs to be used if only one line is available, see Fig. 2.2.8(h). To allow the members to hang vertically during setting, Fig. 2.5.2.4(c), lifting inserts should be installed in the top near the center of gravity instead of using loops in the back. Longer panels may require turning diagrams from the precaster and additional rigging and rigging changes.

Hollow-core or solid wall panels may be placed over dowels projecting out of foundations and then set to bear on shims for elevation control. **Bracing or temporary connections shall be installed and used to plumb the panels prior to unhooking from the crane.** The base is then drypacked and grout is poured into the voids through openings. One alternative for erecting hollow-core or solid slab panels is to cast the foundation wall with a slot to receive the wall panels. The panels are set on shims as required and the slot, which is 2 to 4 in. (50 to 100 mm) wider than the wall, is filled with grout. **Bracing or temporary connections shall be installed and used to plumb the panels.** Differential bowing or camber should be adjusted-out by using connectors at the joints. **Bracing or temporary connections shall not be removed until stability is achieved.**

2.5.3 Floor or Roof Members

2.5.3.1 General. Where several precast concrete members are placed side-by-side in a floor or roof there will often be a unevenness of the top and bottom surfaces due to normal camber and thickness variations. Field adjustments by jacking, loading or other similar procedures can be made to reduce the effect of differential cambers or variations in vertical alignment of adjacent units, prior to the field connection being made between adjacent units.

The architect/engineer generally shows all openings to be provided by the precaster on the structural and/or architectural contract drawings. The GC/CM is responsible for transmitting the shop drawings to the various subcontractors to coordinate the size, shape and location of all required openings.

It is the responsibility of the party that created the opening to provide and install a properly marked and secured hole cover. Openings created in the plant should be covered in the plant. **Openings larger than 2 in. (50 mm) created in the field through floor or roof decks shall be covered immediately.**

Openings may be provided in precast concrete decks by: (1) saw-cutting after the deck is installed and grouted, (2) forming (blocking-out) or sawing in the plant, or (3) using short units framed into full-length units with steel headers or other connections. In hollow-core or solid slabs, structural capacity is least affected by orienting the longest dimension of an opening parallel to the span, or by coring small holes to cut the fewest strands. **Openings in stemmed members shall not cut through the stem. Therefore, the maximum width of any opening shall be narrower than the distance between stems without supplemental structural support.**

Square and rectangular openings larger than 10 in. (25.4 cm) generally are provided by the precaster and are shown on the approved shop drawings. Openings smaller than 10 in. (25.4 cm) are usually cut by the respective trades. A small, hand-held saw with a masonry or diamond blade is often used for cutting these square and rectangular openings.

Openings not shown on the approved shop drawings shall not be cut without written permission of the precaster.

When positioning precast members on a slope, the slings should be adjusted so the unit will hang parallel to the slope. Note that the load will increase on the lower sling.

2.5.3.2 Hollow-Core Slabs. Before erecting any slabs, the supporting structure shall be checked to determine if all bearing dimensions correspond with those shown on the approved erection drawings. Prior to placing the slabs, the bearing surfaces shall be checked to ensure that they are clean, smooth, level and at proper grade. Bond beams should be properly filled, troweled flat and adequately cured. **Bearing material shall be placed as required by the erection drawings.** Care should be exercised to prevent the strips from moving out of position while erecting the slabs. Block walls, steel columns, or other products should not be installed above the hollow-core bearing elevation prior to erection.

In a bearing-wall building, if required by design, the voids at the ends of slabs are generally blocked with commercial shaped cardboard or plastic void fillers, fiberglass wool, stiff mortar, or other suitable materials to prevent grout from flowing into the voids as the end joints are being grouted. The filler material should be placed far enough within the void so that the grout will fill that portion of the slab under the bearing wall.

When hollow-core slabs do not have lifting loops or picking inserts cast in, the slabs are usually lifted with scissor clamps, vacuum lifts, chokers, wire mesh slings or a choker with a spreader bar placed no more than 24 in. (61 cm) from each end, Figs. 2.5.3.2(a) through (f). Otherwise, lifting loops are generally used. Nylon web slings are not recommended due to potential tearing caused by sharp edges.

Slabs without lifting loops cannot be set in final position immediately but will have to be shifted or slid after slings or clamps are removed. This may be accomplished by setting the next slab down and carefully prying against it, or by using come-a-longs, to pull slabs into position, see Fig. 2.5.3.2(b). **Each plank shall have at least the minimum amount of bearing shown on the erection drawings.**

After all units are set in an area, they can be adjusted, using a pry bar, and opened or closed to properly cover the space allotted. **All joints shall be uniform in size.** If the final joint width is more than 1/8 in. (3 mm), it may be necessary to tape the joint or wedge a filler rod into the bottom of the gap to keep grout from leaking through the deck. If a filler rod is used, it should be forced into the very bottom of the keyway gap to permit the grout and keyway to perform properly.

Prior to grouting keyways, the unevenness of tops or bottoms of adjacent slabs should be leveled while keeping the units at proper spacing and at right angles to the bearing wall or support beam. Leveling should be done from the underside, i.e., level the slabs while viewing the evenness across shear keys from below, especially if the deck is to have an exposed bottom. If the deck is to have either tile or carpet applied directly to the top surface, it should be leveled to a smooth continuous top plane. If both surfaces are to be left exposed, it is generally preferred to level the bottom surface then apply a leveling material to the top surface to remove obvious variations. To attain the desired evenness, several methods may be employed such as:

1. Push a low slab upward from below using a leveling screw or pole-mounted hydraulic jack (shores may be installed), see Fig. 2.4.2.5(a).
2. Level with pry bars and drive hardwood wedges or steel shims into the shear key. Projecting wedges should be cut off flush or removed after grout has cured. Care should be taken to maintain uniform joint spacing when using this method.
3. A clamp made with plates top and bottom of shear key with a bolt through plates and joint. Locate clamps at 1/3 or 1/4 points of the shear keys. By tightening bolts, adjacent slabs are forced level. The plates and bolt are removed after grout has cured.
4. Pre-loading the higher slab until the grout is cured.

Connections between slabs over bearing walls are normally accomplished by placing reinforcing bars in the joints between slabs across the bearing walls. Additional bars may be placed into the hollow cores across the bearing wall and grouted. The amount of reinforcement depends on the design requirements and should be specified by the designer and shown on the erection drawing.

For all grouting operations, the grout key shall be cleaned of debris and moistened. Pre-wetting is especially important in hot weather. A grout with a 1:3 cement-to-sand ratio and sufficient water to spread easily should be placed by pumping, wheelbarrows, etc. and spread with brooms, shovels, and/or squeegees over the joints in the deck. **All shear keys between the hollow-core slabs shall be filled, see Fig. 2.4.2.5(c).** The use of high-range water-reducing admixtures in the grout mixture provides fluidity and maintains necessary strength. After grouting the joints, any material that may have seeped through to the ceiling

(a) Use of vacuum lifter.

(b) Use of clamp.

(c) Use of chokers.

(d) Use of four strand lifting loops cast into slab.

(e) Use of clamp and spreader bar.

(f) Use of metal mesh sling.

Fig. 2.5.3.2 Lifting arrangements for hollow-core slabs.

below should be removed before it hardens. Once the grout in the joints has gained strength, the wedges and/or jacking posts may be removed. If necessary, drypack mortar may be tamped into any open spaces. The deck may now be used by other trades. The GC/CM is responsible to ensure that other trades, such as masonry subcontractors, do not overload the deck.

Large openings should not be cut until the joints are grouted and the grout has reached 85% of its design strength.

Plumbers and electricians can easily drill small openings in the field with precaster approval, which may require that the deck be grouted, and if necessary, topped prior to drilling. It is the responsibility of the respective trade to ***always*** cover openings larger than 2 in. (50 mm) to avoid injury.

To produce a smooth, level finish on the top surface as may be required for the placement of tile or carpet, leveling materials such as latex-based cement mortar or a gypsum leveling are sometimes used. These may be poured on the surface usually after grinding to remove excess grout and any other burrs which might project through the skim coat. Some of these materials are self-leveling while others must be trowelled smooth. A concrete topping slab may also be used to prepare a level surface. **The plank shall be grouted before the leveling material or concrete topping is placed.**

2.5.3.3 Stemmed Members. Before erecting double or single tee members, the supporting structure shall be inspected to determine if all pertinent dimensions correspond with those shown on the approved erection drawings. Prior to placing tees, a visual inspection of the bearing surfaces shall be made to ensure that they are smooth and level. The precaster's instructions for handling shall be followed, particularly for long-span, deep-leg members.

Single tees shall be handled in an upright position to prevent buckling and rolling. Single tees can be stabilized during lifting by having leveling loops placed in the flange edges in line with the main lifting loops so that a come-a-long can be placed on the low side to keep the unit level for placing. **Because of their instability, single tees shall be braced to prevent tipping, due to bumping or wind loading, until permanently connected in the structure. Unsymmetrical single tee units shall be maintained in a stable condition until final connections are completed.**

When long tee units are to be used and the tee cannot be rotated, the orientation of the tee on the truck shall be considered.

Bearing pads are normally required between tee stems and bearing surfaces and can be attached to the tee stems prior to hoisting. This method assures proper pad location in relationship to the end of the member. Because tees are not easily moved after having been placed, it is advisable to measure each end of the first unit in a bay and then monitor subsequent units for gain or loss of cumulative dimension, making adjustments as the work progresses.

Welded connections between adjacent deck members are critical to the stability of the structure during the erection process. Cumulative tolerances should be checked and the units adjusted where necessary. **Unless stability of the structure is otherwise provided, sufficient welded connections between adjacent tee units and connections between tees and supporting beams shall be made during or immediately after the erection of a bay.**

Tees normally have bearing plates cast in the stems. Where the erection drawings require the tee stems to be welded, care should be taken to weld only the locations indicated (both ends of the same stem are usually not welded). The connections at one end of the stemmed unit should accommodate thermal movements and rotation. If this is not accomodated, the restraint produced could eventually cause structural damage due to loading, shrinkage, creep and thermal effects.

Flanges shall be connected by using edge connectors welded at locations shown on the erection drawings. Care should be taken in making camber corrections so that flange connectors are not over-stressed causing spalling of the flanges. **Significant differential camber corrections, if necessary, shall be corrected prior to welding flange connectors. In pretopped tees, erection hardware shall be installed below the top surface of the tee to allow for uninterrupted sealant installation.** A range of steel bar sizes should be kept on the site to use as welding slugs or fillers in the gap between the plates. Care should be taken to limit the amount of heat during the welding operation, particularly at low ambient temperatures, as expansion due to heat may cause cracking of the concrete at the ends of the plate. Shimming of tee stems can achieve better camber match between adjacent long- and short-span tees or ramp tees. Uniform joint widths may be important, particularly for exposed ceilings.

After units are set in their final positions, lifting loops shall be removed flush with the deck surface or below the surface if pretopped.

2.5.3.4 Flat Slabs. The handling and erection procedures for flat slabs are similar to those for hollow-core slabs and stemmed members. However, since flat slabs are designed in various widths, lengths and thicknesses and can be prestressed or precast, the precaster should show on the erection drawings any special handling required.

2.5.3.5 Joists. Joists shall always be handled in an upright position to prevent buckling. The precaster's

instructions for handling shall be followed. Long-span joists may need to be handled with spreader bars to achieve vertical pickup at or near each end.

The overall stability of the supporting structure, usually light columns and beams, should be checked and approved by the engineer of record prior to the start of joist erection.

The ends of joists are normally dapped to achieve a hanging connection for immediate stability. One end of the joist is normally welded, while the other end should remain free to allow for thermal movement, rotation and creep. A suitable bearing material should be used to allow the expansion end to move freely. Long-span joists require temporary bracing and guying prior to the application of roof decking. **The method for bracing and guying joists shall be approved by a competent person.**

2.5.4 Cladding

2.5.4.1 General. Non-loadbearing architectural precast concrete and glass fiber reinforced concrete (GFRC) cladding units include items such as wall panels, window wall units, spandrels, column covers, mullions, and soffit units.

GFRC units are usually about 1/4 to 1/5 the weight of similar conventional precast concrete units. Thus, GFRC units permit use of lighter and less expensive handling equipment and connection hardware. A simple hoist, mounted on the roof, or a small crane may be sufficient. However, more care is required during lifting since the lightweight GFRC units are more susceptible to movement due to the wind during hoisting.

Also, when cladding is supported at the ends of cantilevered structural components of the building, the erector should be aware that deflection and rotation of the support, along with the effects of creep, make alignment difficult. Connections should allow for final adjustment at a later time.

While structural steel frames tend to have good dimensional control, rotation and deflection of the steel support beams are much more common than for concrete framing. The erector should be aware that deformation of the supporting structure may induce unintended loading on the cladding unless connections are designed and installed to permit these deformations to occur freely. **The erector shall notify the general contractor and precast or design engineer if excessive deflection and rotation occur.** Connections for cladding should be designed to transmit the vertical loads to the centerline of the steel beam, or provisions made to minimize torsional stresses by specifying heavier members, braces or gussets. It is also desirable, when possible, to concentrate loads at or near the column centerlines.

The installation or prewelding of miscellaneous steel hardware should be performed prior to the start of erection, whenever possible, to minimize the cost of crew and equipment standby.

Inserts for lifting and bracing should be filled immediately after units are erected and aligned if there is the possibility of staining from corrosion or freezing of water in them.

The connections should allow for easy adjustment. Units should be installed to keep loading balanced on the structure or the designer of the structural frame should determine the degree of unbalanced loading permitted with the sequence shown on the erection drawings. Limitations may state, for example, that loading of the structure should be balanced, requiring that no elevation be erected more than a stated number of floors ahead of the remaining elevations; or limitations may involve the rigidity of the structure, requiring that walls should not be erected prior to completion of floors designed to transmit lateral loads.

Units should be aligned to predetermined offset lines, established for each floor level. This is important due to the drift of a high-rise structure caused by sun, wind and eccentric loads and forces due to construction activity being performed by others.

2.5.4.2 Wall Panels. Wall panels should be rigged and hoisted near final location and held in-place until safely secured. Final alignment and connections may be made by a follow-up crew, depending on type of connections used.

It should be communicated to the erector where final connections to steel structures should be completed, and, for cast-in-place structures, what the minimum concrete strength should be prior to precast concrete erection. Consideration also should be given to the number of floors to be erected in a structure on temporary connections, before making final connections.

When wall panels are to be set back into the face of the building, under an overhanging structure, the panels cannot be directly hoisted into place as the structure interferes with the crane lines. Temporary openings in floors above or special equipment, such as a balance beam with a moveable or fixed counterweight, may be used to erect the panels without causing damage or defacing the exposed surface, Fig. 2.5.4.2(a) through (c).

2.5.4.3 Spandrels. Precast spandrels usually extend from column line-to-column line at the building perimeter. They connect either to the columns and/or to the perimeter beams. Precautions should be taken against torsional rotation, due to eccentric loadings on perimeter beams, until final connections are made. Shoring may be required to assist in erection until connections are made.

The erector shall correct misalignment of the units resulting from predetermined deflection and rotation of the supporting structure due to the dead load of the precast concrete units. Units may be erected with a predetermined allowance for such movements or, following field installation, adjustments may be made to connections which are specifically designed for such movement. **Field modifications of connections shall be reviewed and approved by the appropriate designer. Particular care shall be taken to prevent damage to the precast concrete unit when adjustments are being made to bring the unit into final position.**

For long slender spandrels, a spreader bar can be used when hoisting to minimize rigging requirements and to reduce the forces which could cause bowing or buckling of the member.

Spandrels shall be safely secured, prior to releasing the load and disconnecting the rigging. For long slender spandrels, care should be taken to ensure that the spandrel is secured with a temporary connection until a slab-tie connection is made or is secured with a permanent connection.

Spandrels should be aligned to predetermined offset lines established for each floor level. The erector should understand the glazing interface details and tolerances. Vertical dimensions between spandrels should be checked with a "story pole" or similar device to ensure that opening size is within allowable tolerances.

2.5.4.4 Column Covers and Mullions. Column covers are usually manufactured in single story units and extend either from floor-to-floor or between spandrels. However, units two or more stories in height may be used and care should be taken when hoisting and rotating them to prevent cracking. Two lines may be required to rotate the member.

When erecting column covers and mullions, slings are attached to lifting devices at the top. The units are then rotated to the vertical position, Fig. 2.5.4.4. After rotation the unit is hoisted and placed onto the structure near its final position. **Column covers shall be safely secured prior to releasing the load and disconnecting the rigging.**

Column covers and spandrels are often erected sequentially. Caution should be taken so that one member does not cause stress on the other. Temporary braces may be required until the spandrel is erected above the column cover and permanent attachments made.

When column covers and mullions are interspaced with windows, the erector should understand the glazing interface details and tolerances.

(a) Granite-faced precast with shipping chairs. Chairs are also used as part of a special erection device. Erection cantilever device installed on 17-ton granite-faced precast unit allows panels to be erected under overhanging structural slab. Note location of the crane lines in front of precast unit.

(b) A special placement device designed to attach to the decorative face of precast concrete wall panels, lift and level them from trailer easel and position them in the building at a cantilever distance of 4 ft. from the crane hook centerline. A moveable counterweight that rolls back and forth in excess of 12 ft. compensates for the weight of the panel and allows end to end levelling.

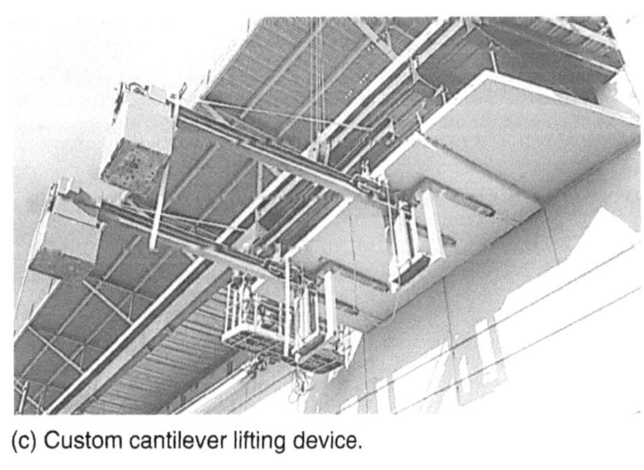

(c) Custom cantilever lifting device.

Fig. 2.5.4.2(a) through (c) Use of special equipment to erect panels under an overhanging structure.

Fig. 2.5.4.4 Column cover is rotated to vertical position using inserts on top of units.

2.5.4.5 Soffits. Soffit units are normally erected under perimeter beams to form an architectural enclosure. To achieve this, special erection equipment is usually required.

One method of erecting precast soffits is to attach slings to lifting devices and the soffit hoisted onto cantilever shoring and disconnected. The soffit unit is then rolled under the beam and jacked or hoisted into final position. Connections are made and shoring removed. Other methods of placement are cantilever spreader bar, scissor lift and genie hoist. All of these methods can be costly and very time consuming and require a great deal of preplanning and preparation.

2.5.5 Bridge Members[14]

2.5.5.1 General. The method chosen to erect precast concrete bridge members depends primarily on member weights and lengths, crane capacities and access conditions at the site. Most bridge members are long, heavy sections which require chokers, slings, hooks and shackles appropriately sized to safely accommodate the loads being hoisted. Bridge erection is generally unique and the erection sequence and fall protection method to be implemented should be evaluated and thoroughly planned.

Bridge member type varies based on loading and maximum span requirements. These requirements, along with aesthetics, clearances, use and economics, have led to numerous precast sections being used for bridge members such as double-tees, I-beams, bulb-tees, box beams, voided slab beams, solid slabs and various types of hollow segmental sections.

Bridge members may range from short spans [10 to 70 ft (3 to 21 m)] to long spans up to 180 ft (55 m). Depth of bridge member, combined with spacing, control the span length of bridge members. Shorter segmental sections can be post-tensioned together to create spans in excess of 200 ft (61 m).

The preferred method of erecting long beams is with a single crane located at either bridge beam support or somewhere between supports. Single cranes located at a support are geneally limited to short spans of 60 ft (18.3 m) or less. I-beams as long as 120 ft (36.6 m) have been erected with a single crane placed at midspan. This requires good access for both the crane and the delivery vehicle near midspan to reduce the reach the crane must make to lift the beam, Fig. 2.5.5.1(a). Single-crane lifts require a sufficient length of boom to keep the cables at a specified minimum angle from horizontal, generally 60 deg. Spreader bars or struts can also be used to maintain this minimum angle.

Erection utilizing two cranes is normally used when long beams can be delivered along the bridge span. Individual cranes positioned at the supports can swing the beams from the delivery vehicle directly to their final position, Fig. 2.5.5.1(b). Dual-crane lifts do not require the very long boom lengths of single-crane lifts. This is important in situations where headroom is limited, such as erection under a bridge overpass or near or under power transmission lines.

Passing beams from crane-to-crane is normally required when erecting long beams over waterways and railways, where neither cranes nor delivery vehicles have access between supports, Fig. 2.5.5.1(c).

Launching trusses are used when, due to ecological or practical considerations, the methods described above are not feasible. This generally occurs in areas with limited access or for long spans in the range of 160 to 200 ft (48.8

Fig. 2.5.5.1(a) Single crane erecting long beam.

Fig. 2.5.5.1(b) Dual-crane lift of long beam.

to 61 m). Launching trusses eliminate the need for cranes, delivery vehicle or temporary support towers to be placed near midspan, and can be used for both single-piece or segmental construction. Launching trusses are usually shipped in pieces and assembled at the jobsite. Methods of installing the trusses vary depending on the type of construction.

The bearing supports are important to provide uniform bearing for the generally high concentrated forces at the beam-ends. Elastomeric or laminated fabric bearing pads are preferred beam supports.

Fig. 2.5.5.1(c) Two cranes transfer the weight of the far end of this beam using a triangular transfer plate assembly.

Prior to mobilizing erection equipment, the support surfaces should be checked for horizontal and vertical control, as well as flatness and level or slope.

When precast concrete beams are too long or too heavy to be shipped as a single piece, they can be cast in segments, erected on temporary support towers, and spliced together in their final positon, Fig. 2.5.5.1(d). Support towers usually extend the full width of the bridge to accomodate continuous erection and splicing operations. A solid foundation, usually a compacted crushed gravel base, must be provided for the towers, since very little settlement can be tolerated between the time the splice is completed and the post-tensioning is applied across the joint. Two or more timber mats, placed in perpendicular directions, support the towers and distribute the loads to the base.

It is important for the erector to install two lines of girders with diaphragms or "stiffeners" attached. Diaphragms are normal to the longitudinal axis of the bridge and connect precast flexural members to one another in order to achieve stability and prevent beam overturning.

Diaphragms can be fabricated as separate precast pieces and shipped loose to the jobsite for installation into the structure, Fig. 2.5.5.1(e). Steel sections or pre-assembled frames are also used as diaphrams.

Fig. 2.5.5.1(d) Beams supported on temporary tower. (Note braces used to stabilize beams.)

After the beams have been erected, and before they are permanently connected into the structure, they can be subjected to forces of nature that could cause them to fall off the supports. Forces include wind, earthquake, or thermally-induced sweep. Temporary braces consisting of steel or timber should be used to stabilize the beams. Braces are removed after the final connections are made, see Fig. 2.5.5.1(d).

2.5.5.2 Deck Panels. Precast composite bridge deck panels are usually 3 to 4 in. (75 to 100 mm) thick concrete slabs which span between the top flanges of concrete or steel beams, Fig. 2.5.5.2. These panels provide a working platform for deck reinforcement placement, and a composite stay-in-place form for a cast-in-place concrete overlay. Bridge deck panels can be handled by conventional techniques or with proprietary lifting equipment.

2.6 Post-Installation Considerations

2.6.1 Protection of Work

All precast concrete should be furnished to the jobsite in a clean and acceptable condition with embedded plates cleaned and inserts free of obstructions and kept in such condition until erected. The erector should take necessary

Fig. 2.5.5.1(e) Fabricated diaphragm being installed into structure.

Fig. 2.5.5.2 Deck panels spanning between the top flanges of concrete and steel beams.

precautions to protect the work and materials of other trades from damage during erection.

At the end of each working day, measures may need to be taken to protect the installation from weather damage. For example, adequate temporary protection should be provided against freezing of water in holes, pipe sleeves and inserts.

The erector should protect adjacent materials, such as glass and aluminum, from damage by field welding or torch cutting operations. Therefore, the sequencing of work of other trades should be taken into consideration by the GC/CM to prevent such damage.

The GC/CM should provide and maintain temporary protection to prevent damage or staining of exposed precast concrete during construction operations after installation. Rainwater or water from hoses used during the construction of the building, can cause discoloration of the precast concrete units by first washing across other building materials (such as steel, concrete, or wood) and then across the precast. Particular care should be taken to prevent this from occuring. Dirt, mortar, plaster, grout, fireproofing, and debris from concrete placement should not be permitted to remain on the precast concrete. **When the**

erector is aware of conditions that could result in damage to or staining of erected precast, the erector shall notify the GC/CM and/or precaster.

After final erection of any portion of precast work to acceptable alignment and appearance, including completion of all connections and joints, the general contractor should assume responsibility for protection of the work. Any cleaning or repair of precast concrete work subsequent to installation and/or acceptance is the responsibility of the general contractor. Specifications should state this responsibility clearly. It is entirely impractical for the precaster or erector to police the work against damage by others after the precast has been put in place. There should be a carefully established and implemented program of protection and later cleaning for each job, under the responsibility of the general contractor, who alone can control all of the potential sources of damage.

2.6.2 Repairs at the Jobsite

All lifting and handling devices cast into the members should be dealt with as specified on the erection drawings or based on one or more of the following conditions:

1. Removed where they interfere with other trades.
2. Removed and surface patched where demanded by appearance requirements.
3. Removed or protected from possible corrosion and marring of the finished product where not exposed to view.
4. No action required.

When the erector performs the patching and if it is necessary to provide lifting and handling devices in finished edges or exposed surfaces to allow for attachment of temporary lifting and handling devices, these holes shall be patched. Often, these devices may be recessed, filled, and finished at a later date with prior designer approval. However, specialized lifting equipment may be necessary to eliminate the patching of exposed lifting or handling devices.

Immediately after erection, the erection supervisor or precaster's representative should make a thorough inspection of the precast concrete installation and arrange for final repairs, cleaning where needed, joint treatment and final acceptance by the architect/engineer. The precast elements may sustain superficial damage (minor chipping and spalling) during handling, transportation or erection that will require jobsite repair. **Major repairs shall not be attempted until an engineering evaluation is made to determine whether the unit will be structurally sound. The repair method to be used shall be approved by the precast engineer.**

Repair and patching of precast concrete is both an art and skill requiring expert craftsmanship, if the end result is to be structurally sound, durable and pleasing in appearance. Responsibility for repair work is normally resolved between the precaster and the erector. The decision of when to perform the repairs should be made by the precaster, who should be responsible for satisfactory final appearance. **When the erector performs the repairs or patching, the erector shall inspect them to see if they are aesthetically acceptable. In this case, the erector shall ensure that all repairs and patching are completed.**

Since the techniques and materials for repairing precast concrete are affected by a variety of factors including mix ingredients, final finish, size and location of damaged area, temperature conditions, age of member, surface texture, etc., precise methods of repairing or patching cannot be detailed in this Manual (see Appendix E for guidance on techniques and materials).

2.6.3 Cleaning

Final cleaning should be performed after joint caulking (see Appendix F for information on caulking) is completed and no earlier than three days after any repairs have been completed. See Appendix G for cleaning procedures.

If possible, cleaning of the precast concrete should be done when the temperature and humidity allow rapid drying. Slow drying increases the possibility of recurring efflorescence and discoloration.

A small [at least one square yard (0.8 sq. meters)] inconspicuous area should be cleaned and checked to be certain that there is no adverse effect on the concrete surface finish or adjacent materials prior to proceeding with final cleaning. The effectiveness of the method on the sample area should not be judged until the surface has dried. Materials such as glass, metal, wood, stone, or concrete adjacent to the area to be cleaned should be adequately protected since they can be damaged by contact with some stain removers and cleaning agents or by physical cleaning methods. A sprayed-on, strip-off plastic can be used to protect glass and aluminum members.

2.6.4 Acceptance

Specifications should spell-out acceptance procedures and the contract documents should identify who the accepting authority will be — GC/CM, architect, engineer, owner or jobsite inspector. *One person should have final and undisputed authority.*

Upon substantial completion of the precast work, the GC/CM together with the architect/engineer and/or accepting authority should make careful inspections to ensure the correction of any deficiencies.

The erector/precaster should periodically request approval and acceptance for all precast concrete work completed, as deemed necessary, in order to adequately protect the interests of everyone involved in the project. In most cases, the size and nature of the project will dictate the proper intervals for securing approval and acceptance. Periodic approval in writing should be considered when it appears that such action will minimize possible problems which could seriously affect the progress of the project.

Upon notification by the erector/precaster, the GC/CM and the architect and/or engineer should promptly make arrangements for the final inspection of the erected portion of the work to determine if it is plumb, level, aligned within the established tolerances, and that the proper connections have been made.

Following the final inspection of the erector's work, the architect/engineer should immediately prepare a list, setting forth in accurate detail any items of the work that are found not in conformance with the contract documents so that proper corrective action can be taken in a timely manner.

Following the preparation of this list, a meeting between the contractor, the precast concrete manufacturer, the erector, and the architect/engineer should be held promptly to discuss any questions concerning what the architect/engineer requires to be done before the work can be accepted as complete. Units found to be slightly damaged may be authorized for repair at this time.

When advised by the erector that the listed items have been completed, the GC/CM and architect/engineer should check the corrections. After the precast units have been accepted, subsequent responsibility and liability rest with the GC/CM and the architect/engineer.

DIVISION 3 — EQUIPMENT

3.1 General

Erection supervisors are not generally expected to know how to operate the equipment they supervise. However, they should be familiar with the proper set-up, limitations, and safe use of the equipment. They should also obtain the operator's inspection report prior to giving instructions to proceed with the work.

3.2 Operator Training and Qualification

The safe and efficient use of cranes, hoists, personnel lifts and other specialized equipment requires operator training and qualification. **The erector shall have competent and careful operators who are physically and mentally fit, thoroughly trained in the handling of the loads and the safe operation of the specific equipment they will be expected to operate.** Serious hazards can result from improper hoisting operations or by using equipment for a purpose for which it was not intended or designed. Failure to review operational and maintenance instructions for each piece of equipment and to identify the operational differences can cause accidents even when the equipment is operated by trained operators.

The equipment operator shall have knowledge and understanding of equipment operating characteristics, capabilities, and limitations including: equipment-rated capacity and effect of variables on capacity, safety features, required operating procedures, and shall demonstrate skill in manipulation and control of equipment through all phases of operation.

Erectors who employ personnel who operate equipment covered in this Manual shall assure themselves that the personnel are competent to perform the operations. Every equipment operator shall know the following, if it relates to their specific equipment:

1. **Equipment operator's responsibilities.**
2. **Inspections and routine maintenance requirements of the specific equipment.**
3. **Basic requirements for transporting the equipment to and from the jobsite.**
4. **Pre-start-up requirements.**
5. **Set-up requirements, such as requirements for use and placement of outriggers.**
6. **Operational requirements, working range, swing radius, and limitations of the equipment.**
7. **How to operate the equipment and what to do in an emergency.**
8. **Appropriate hand signals and verbal directions.**
9. **Specific equipment hazards, such as pinch points and blind spots.**
10. **Load limits and how to read load-limit charts.**
11. **Understand set-up programming and use of onboard computer.**
12. **Proper rigging requirements and the safety requirements for the use of slings and related equipment.**
13. **How to safely maneuver the equipment on the jobsite, shut-down, park, and secure the equipment.**
14. **OSHA regulations and company rules for equipment operation.**
15. **Specific jobsite conditions, such as the location of curbs, sidewalks, buried gas lines, electric cables, underground utilities, and the location of overhead power lines and the minimum federal, state, and local clearances required.**

3.3 Selection of Handling Equipment

Considering the many variations in crane and special hoisting equipment, this section is not meant to cover all of the available equipment, but rather the most commonly utilized equipment in the precast concrete industry.[15,16]

An analysis of each job shall be made to select the most appropriate crane and/or other erection equipment, and to select the optimum site locations from which the greatest number of units can be safely handled without equipment movements.

When selecting equipment, it is necessary to ensure that it will be safe, reliable and in good working order in all anticipated conditions to which it will be exposed and operated.

All wire ropes for hoisting equipment shall be of adequate size, and properly maintained, not kinked,

twisted, torn, or stretched. Wire rope shall be replaced if wear or damage is evident in accordance with recommendations in PCI "Erection Safety Manual for Precast and Prestressed Concrete" (MNL-132).[17]

All lifting beams, spreader bars, or lifting trusses shall have their lifting capacity plainly marked on each unit. The ultimate capacity shall have a minimum safety factor of 5.

Crane selection is affected both by where the precast unit must be placed and from where it must be lifted. Many cranes are rated differently when lifting over the side compared to the rear. Some have no rated capacity over the front. **For large or long precast shapes, careful consideration shall be given to where the crane and delivery equipment are located and oriented.**

The crane selected should meet or exceed all job requirements and have capacity in reserve (including rigging) for unanticipated problems with respect to the load capacity of every lift. The following list of considerations is a guide. Specific jobs, cranes, and rigging considerations should also be taken into consideration.

A crane shall be selected considering the following criteria:

1. **Maximum pick — load versus reach — (often not the heaviest piece) needs to be determined.**
2. **Size and shape of structures or obstructions which may effect boom length.**
3. **Weights, dimensions and lift radius of the heaviest and largest pieces. (The correct weight is determined by adding the weight of the piece and the weight of all rigging hanging from the boom tip which include weight of jibs, hooks, blocks, slings and other load handling devices including the hoist rope below the boom tip), Fig. 3.3(a).**
4. **Maximum and minimum lift height and radius, and the weight to be handled at each elevation location.**
5. Number of lifts that have to be made and with what frequency.
6. **Specific crane charts for the pieces of equipment being considered need to be reviewed so that all parameters are met, i.e., boom, counterweights, etc., Fig. 3.3(b).**
7. Mobility required due to site conditions and access for trucks and cranes.
8. Length of time the precast members must be held aloft (hydraulic systems tend to creep).
9. **Loads that must be walked or carried.**
10. Availability and cost of equipment.

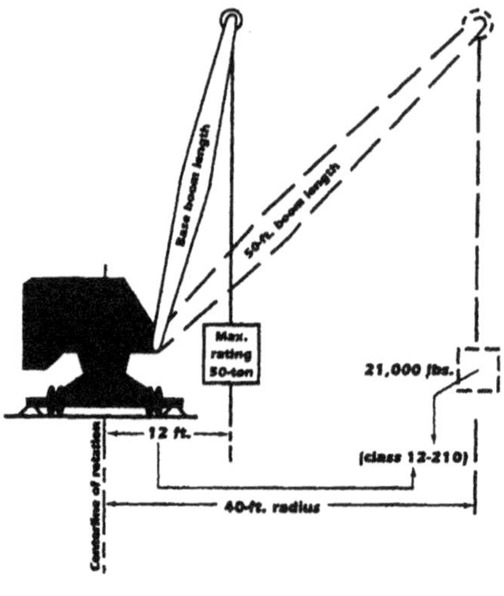

A 50-ton, class 12-210, truck-mounted crane. But how do those numbers relate to what the crane can do?

A 50-ton crane can lift 50 tons, but not with any length of boom or at any distance from the centerline of rotation. Class 12-210 indicates the crane can lift 50 tons at a 12-foot working radius. The 210 represents the load, in hundreds (210 x 100 = 21,000 pounds) that can be lifted safely at a working radius of 40 feet.

The lifting capacity at a 40-foot working radius is based on a crane operating with a 50-foot boom. Cranes equipped with larger boom lengths have a lower lifting capacity because of the increased overturning load from the heavier boom.

A crane rating chart, located in the cab, is accessible to the operator. This chart provides the lifting capacity of the crane for different boom lengths and working radii.

Fig. 3.3(a) Considerations for crane capacity.[18]

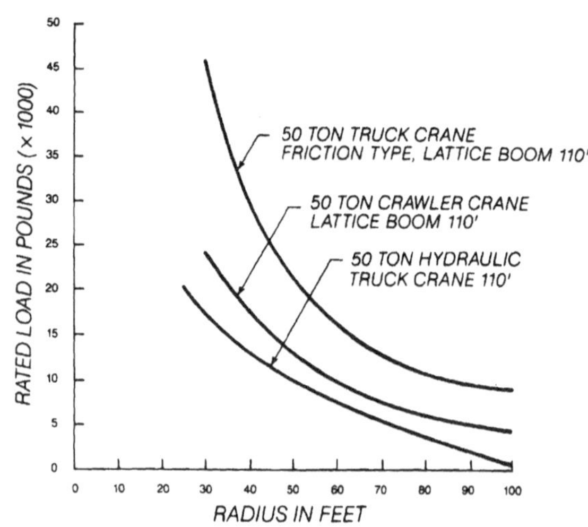

Fig. 3.3(b) Rated lifting capacities — Load vs. Radius. **(This is for illustration only. Specific manufacturer's crane chart shall be used.)**

11. Cost of mobilization, on and off-site transportation, and altering boom length.

12. Space large enough to assemble and disassemble the crane.

13. Clearance between the load and boom and head room between the load and rigging required to make the lift as well as overhead obstructions to avoid two-blocking.

14. The following considerations shall be evaluated when turning (rotating or tripping) precast units in the air using both the main and auxiliary (whip) lines:

   (a) Verify that whip has sufficient line pull. Remember that it may initially carry more than half the weight of the unit.

   (b) Verify that the crane is capable of multi-parting the whip as well as the main load line.

   (c) Consideration should be given to restraining the load from spinning while lifting with two (2) lines; i.e., if the main and auxiliary lines go over adjacent sheaves so that the lines are side by side, the piece will also try to orient itself to match. It is important to be ready for this possibility. (Refer to Section 2.2.8)

Placing the crane next to the building puts lateral pressures on the building's foundation walls. As shown in Fig. 3.3(c), distance from the crane to the building should be equal to or greater than the depth of the footing.

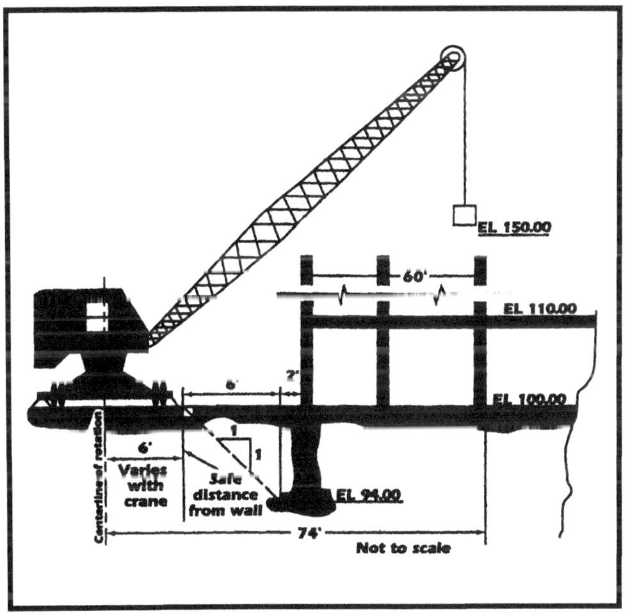

Fig. 3.3(c) This typical precast building representation demonstrates how to calculate the necessary dimensions for crane selection. When locating the crane be sure to place it a safe distance from the building to avoid increasing lateral soil pressures on the foundation walls.[18]

3.4 Safety Devices

The weight of the load to be lifted or the force to be exerted by a crane shall be determined to assure the equipment and its attachments are not overloaded.

Equipment is frequently equipped with load indicating computers and other electronic and/or mechanical safety devices to warn against hazardous conditions. Cranes, for example, are frequently equipped with anti-two-block devices and limit switches that prevent contact between the load block or ball and the boom tip. These devices, often installed by new equipment manufacturers, can be retrofitted to older cranes. **Anti-two-block devices shall be provided on all telescoping boom cranes and lattice boom cranes when using suspended personnel platforms.**

Cranes can also be equipped with a computer and electronic load moment indicators that display conditions relating to crane safety and operation. These devices display the weight of the load on the hook, safe working load, load radius, boom angle, and allowable safe working loads based on the boom length and height. Some of these units are equipped with visual and audible alarms and automatic cutoffs that signal when the crane is approaching an overload situation.

Manufacturers of electronic load measuring devices and dynamometers have simplified these tasks by offering small, relatively inexpensive measuring devices that can be attached to slings and crane cables. These electro-mechanical devices improve safety and reliability.

3.5 Cranes

3.5.1 Mobile Cranes

3.5.1.1 Conventional Cranes. Truck- [Fig. 3.5.1.1(a)] and crawler-mounted [Fig. 3.5.1.1(b)] cranes are most commonly used to erect precast units. Lifting capacity can be governed by the structural capacity of the crane or by stability against overturning. **Load capacity charts shall be strictly adhered to.**

The general characteristics of the truck-mounted conventional crane are as follows:

1. Travels over roads.

2. Has excellent capacity when working over rear; however, decrease of capacity when working over sides and front.

3. Possesses a wide range of boom and jib length combinations, and also a wide range of capacities for the various cranes.

4. For most large cranes, the boom and counter-weight must be assembled on the jobsite.

Fig. 3.5.1.1(a) Truck-mounted lattice boom crane.

Fig. 3.5.1.1(b) Crawler-mounted lattice boom crane.

5. Travels with loads according to the on-rubber charts.

The general characteristics of the crawler-mounted crane are as follows:

1. Must be transported to the job on trucks.
2. On large capacity models, crawlers and superstructure must be assembled on the jobsite.
3. Good jobsite mobility even on soft and/or uneven ground.
4. Can walk with load on stable ground.
5. Good capacity for crane rating (rated through 360 degrees).
6. The boom and counter-weight must be assembled on the jobsite.
7. A wide range of lifting capacities with boom and jib combinations.

In general, a truck-mounted conventional crane will have a greater lifting capacity at all working radii than a crawler crane of the same basic rating. The rubber-tired crane equipped with outriggers can be leveled more accurately and the ground conditions under outriggers can be more readily established. This, coupled with a basically heavier carrier accounts for a better performance curve for the conventional truck-mounted crane. The lower rating for crawlers has been established because there is some uncertainty about the location of the tip axis and, therefore, overturning cannot be calculated as precisely as it can for other types of equipment or outrigger systems. In addition, the reduced rating recognizes the fact that these machines are intended for use on soft ground.

Other considerations which could affect crane selection are:

1. Both truck-mounted and crawler cranes can be equipped with booms of varying lengths by adding boom sections. However the length of each boom section is set by the manufacturer and cannot be modified. This is an important aspect when working under overhangs or within areas restricted in heights.
2. Since the boom is assembled on the site in a horizontal and down position, one should consider the availability of a relatively long, flat and unobstructed area to allow for the installation of the boom sections and jib, if required.
3. When assembling the crane in a restricted area, consider availability of space to allow the crane cab (house) to rotate from the pickup point to the installation zone. The turning radius for a large machine often exceeds 30 ft (9.1 m) and needs to be taken into account relative to the space allocated for the operation of the crane.

3.5.1.2 Telescoping (Hydraulic) Boom Cranes [Fig. 3.5.1.2(a)(b) and (c)]. These cranes have the distinct advantage of a self-erecting boom and can be ready to work minutes after arrival on the jobsite. The telescoping boom feature also provides flexibility for erecting precast members where overhead obstructions occur or access is restricted for other reasons. A disadvantage of the telescoping boom crane is the lower load capacity at increasing working radii as compared to a conventional crane with a pinned lattice boom. This disadvantage is somewhat offset by the fact that the center of rotation is ahead of the boom attachment pins, and thus the crane can often be positioned closer to the work when operating over the rear of the carrier. Also, most of the smaller cranes are not equipped with two winches which are needed to trip (turn) panels. A very important disadvantage of hydraulic cranes is that they are usually not rated through 360 deg. and the capacity over the front of the carrier is substantially reduced or, in some cases, not recommended by the manufacturer. In general, the operating cycle is slightly slower than for a conventional crane. Telescoping boom cranes are available as truck-mounted, carrier-mounted and self-propelled models.

The general characteristics of the telescoping boom crane are as follows:

1. Can travel over the road.

2. Good jobsite mobility on relatively firm ground.

Fig. 3.5.1.2(b) Telescoping (hydraulic) truck crane.

Fig. 3.5.1.2(a) Telescoping (hydraulic) boom crane.

Fig. 3.5.1.2(c) Stationary-mount hydraulic rough terrain crane. (Note: Operator does not swing with the boom.)

3. Can walk with light loads under limited conditions, based on manufacturer's recommendations.

4. Length of boom is variable up to the maximum equipped length.

5. Can work under overhead obstructions.

6. The heavy boom causes the lifting capacity to decrease more rapidly as radius increases.

7. On a confined site, can work on a very tight radius by either booming-up, and/or telescoping-in.

8. Constructed such that the boom is actually behind the center pin.

3.5.1.3 Mobile Luffing Jib Cranes. The luffing jib crane is perhaps the most versatile crane on the market, Fig. 3.5.1.3. The ability to luff the main boom and luff the jib as well, plus a fixed jib provides the erector with reaches over tall structures that previously could only be accomplished by tower cranes. Converting from a conventional jib to a luffing jib requires a special point section and a third drum.

Since much of the load chart is based upon factors other than tipping, the utmost care must be given to ensure that all loads are within the load chart. These cranes require special site preparations in the set-up areas.

3.5.1.4 Mobile Tower Cranes. These cranes often offer more flexibility and more capacity. Many of these cranes are actually conventional mobile cranes with tower attachments. Like mobile cranes, they can be truck-mounted or on crawlers, Fig. 3.5.1.4.

Some advantages are that:

1. They do not require specially engineered foundation or building alterations.

2. They usually cost less to assemble.

3. They are mobile, but do require level firm ground and possibly mats.

4. They can be repaired more easily on the ground.

5. The operator can usually see loads when making a pick.

6. The load does not bounce quite as much.

Some disadvantages are:

1. Rental costs are usually higher.

2. Operator is on the ground and often cannot see the load through most of the lift.

3. There are height limitations.

4. They require more room to assemble and set up.

Fig. 3.5.1.3 Mobile luffing-jib crane.

Fig. 3.5.1.4 Mobile tower crane (track mounted).

3.5.1.5 Stationary Crawler Cranes (Stationary crawler crane in a ringer configuration). Ringer cranes are used to increase the lifting capacity of an otherwise mobile track crane, Fig. 3.5.1.5. The decision to use a stationary crawler crane in a ringer configuration rather than a mobile crawler crane should be made carefully because of the economic ramifications. Among the determining conditions are:

1. Whenever the lift heights are extreme and the reach into the structure may be deep.
2. Whenever the site is constricted or congested and will not allow the use of mobile cranes.
3. Whenever the need for a crane is long-term in a given setup location.

A disadvantage of these cranes is the cost and setup time.

Fig. 3.5.1.5 Ringer crane.

3.5.2 Tower Cranes

When the site is constricted, the schedule relatively long and the building tall, tower cranes should be considered. Tower cranes range from fixed cranes that are secured to the structure or to a footing outside the structure, to mobile tower cranes that can be on a track system or can be assembled at the site and moved around much like a conventional mobile crane. The common feature is that the boom is at the top of the tower. Some cranes are equipped with luffing booms with the load line at the end of the boom and others with a horizontal boom where the load line travels back and forth on that boom. The operator is generally housed at the top of the tower except in the case of mobile tower cranes where the operator is using a remote control. When installed prior to the erection of the structural framework, fixed tower cranes generally require some sort of foundation which may become part of the building structure or be removed or buried at the completion of the job. Thus, because of the cost for erection, dismantling and foundation work, these cranes are suitable for long-term jobs or where, due to limited access or other reasons, no other equipment will be adequate for the precast erection.

The decision to use a stationary tower crane, Figs. 3.5.2(a to c), rather than a mobile crane should be made carefully because of the limited capacity for typical precast loads at longer reaches. Among the determining conditions are:

1. Whenever the need for a crane is long-term in a given location.
2. Whenever the site is constricted or congested.
3. Whenever the lift heights are extreme and the reach into the structure may be deep.
4. Whenever there is little need for mobility but lift frequency and speed is high.
5. Whenever the load placement in the working area is more easily done by a crane working from within that area rather than by one outside.

Some disadvantages of these cranes are as follows:

1. The operator is often in the blind when unloading or setting units and thus operating signals are transmitted by telephone or radio.
2. A single-hoist drum, which may require additional considerations for turning and handling of precast units.
3. Smooth and precise piece control is difficult because of sway and bounce of the crane.
4. Lifting capacities of tower cranes are limited.
5. Wind susceptibility.

Fig. 3.5.2(a) Stationary trolley tower crane.

Fig. 3.5.2(b) Stationary luffing tower crane.

The tower crane may be:

1. Located external to the structure.
   a. Ground-supported at a set location but tied to the structure at intervals.
   b. Rail-mounted units, allowing free horizontal movement along the face or perimeter of the structure.[19]
   c. Trailer-mounted tower cranes.
   d. The type that climbs with the building.

   External cranes are often preferred on building sites for these reasons.

   a. They require minimal special design treatment for floors, although cost of foundations may be substantial as the entire weight of the crane is carried on its tower and transmitted to the base.
   b. Stripping and removing the crane is easier as the crane can lower itself almost to ground level.
   c. The crane operator can usually see the load and the surrounding area, at least during some stage of the hoisting operation.
2. Located within the structure.

   A tower crane located inside the building structure has the following advantages over external tower cranes:

Fig. 3.5.2(c) Platform tower crane.

a. Smaller radius to cover a site.

b. Very little longitudinal bracing required.

c. Not susceptible to as much wind loading.

d. Normally no interference with external wall or cladding.

The location of an internal crane is based on positioning that would involve the least amount of interference with construction trades and be coordinated with structural drawings for beam and column locations. The initial selection of the tower crane position relative to the building geometry is extremely critical. It may be necessary to avoid the elevator shaft location, since elevators require a considerable amount of lead time for final installation. Furthermore, the crane usually is not located in the stair shafts as this area is important for workmen's daily access between floors of the building. Basically, the crane should be located in such a manner as to avoid interference with extensive interior work, while remaining compatible with the structural framing member locations and sizes.[20]

In addition to considering such factors as the weights, dimensions and lift radii of the heaviest and largest loads (at selected and available unloading areas and at point of installation), the erector when selecting the crane for a specific project should also consider the characteristics of the available cranes in relation to:

a. The type, size and location of the base for the crane. This is usually a function of the soil bearing capacity and the type of machine.

b. The maximum free-standing height of the crane.

c. The maximum braced height.

d. The climbing arrangement.

e. The weight of the crane to be supported by the building frame.

f. The jib lengths available considering swing radius obstructions.

g. The available headroom between the maximum height position of the hook with reference to the area of erection and possible obstructions.

h. The area that must be covered. Since tower cranes have limited capacity at extreme reach, precast units may be fed into a special bay or near side from which they are hoisted and swung into final position by other equipment.

i. The hoisting and lowering speeds.

j. The length of cable the hoist drum carries.

k. The number of parts of line the machine needs for the required hoisting.

l. The cost of operations such as erection, dismantling, off-site transport and costs per lift (climb) of the machine.

m. The loads imparted to the building structure.

n. Power sources such as generator and/or temporary electrical supply.

o. The method of crane erection and dismantling.

The erector shall read and understand the load charts. The manufacturer's stated conditions for support, counterbalance and ballast, assembly, erection and limiting wind speeds shall be met. Nearly all tower cranes require ballast or counterweights which are an important consideration during erection, operation and disassembly.

Scheduling problems may be encountered unless firm time allocations for separate trades can be maintained by the general contractor. Some erectors occasionally make use of night-time hoisting with tower cranes, transferring units to block and tackle equipment or setting them temporarily for final placement or alignment during daylight hours.

3.5.3 Monorails

Precast wall panels can also be erected using a monorail system. Basically, a monorail consists of a continuous track or I-beam installed at the perimeter of a building which allows panels to be transported horizontally into position by carrier or hoist, Fig. 3.5.3. This specialized system enables an erector to place panels in areas normally not accessible or within the reach of cranes. The monorail may be fed remotely by a crane or a derrick or can be loaded directly using electronic hoists.[21]

Monorail systems are favored by some erectors for buildings above 16 or 20 stories and where relocation of such rails can be made in jumps of at least 10 to 15 stories.

The erector should recognize that this may be a more costly erection method when planning a project. Furthermore, since the monorail is cantilevered from the building perimeter an analysis of the loads imparted to the supporting structure during the hoisting operation is an absolute requirement. Review by a professional engineer is essential before using this method to erect precast panels. Generally one should only consider using a monorail to erect precast panels after reviewing the following:

1. Are the units to be erected relatively light?

2. Will the existing structure withstand the loads imparted to the supporting structure during the hoisting operation?

Fig. 3.5.3 Chicago boom is used to lift the architectural panel from the ground and transfer it to monorail for actual erection.

3. Is the building geometrically similar throughout the height of the structure? Modifications to a monorail are expensive.

4. Is the site congested, with very limited access negating the use of mobile cranes?

5. From the available unloading area, can direct hoisting to the monorail with electric hoists be accomplished? If not, consider the possibility of using a derrick to an intermediate unloading platform for transfer to the monorail system.

6. Does the erection schedule require more than one set-up? Installation and dismantling costs are considerable and thus should not be overlooked.

7. How does the slow horizontal speed affect erection? Distance from the transfer area to the final position of the precast unit is an important factor when assessing the daily installation rate.

8. Does the installation process affect other trades?

3.5.4 Derricks

A derrick is basically a stationary, anchored boom structure suitable for handling relatively heavy loads, at medium radii. The more familiar derricks consist of a

Fig. 3.5.4(a) Outrigger-mounted guyed derrick. This arrangement permits swinging loads through guy positions by booming the derrick to a close radius, allowing the boom tip to pass under the guys while supporting the load.

Fig. 3.5.4(b) A roof-mounted stiffleg derrick was used to erect the architectural panels on this parking deck.

Fig 3.5.4(c) Closeup of rigging on a Chicago boom. The footblock and top pivot are mounted on the building structure and boom can be rotated through 270° if corner mounted (see Fig. 3.5.3 also).

mast and boom. The mast is supported either as a "stiff-leg" derrick or as a "guy" derrick, Figs. 3.5.4(a) and (b). In either case, the mast and boom rotate freely on a bull wheel. The boom is luffed in much the same fashion as a conventional crane.

In general, a derrick is employed where precast units are placed on or combined with a structural frame and when the job duration is long enough to justify the considerable expenses of setting up and anchoring the hoisting apparatus.

When considering the installation of a derrick for precast erection, the following points should be considered:

1. Can the building structure physically accommodate a derrick?

2. Will the existing structure withstand the imparted loads?

3. Does the erection schedule require installation of precast units prior to the complete erection of the building structural frame work? If this is so, consider use of a Chicago boom-type derrick, since stiffleg and guy derricks cannot be easily installed in "between-floor" conditions unless extensive structural framing modification were made to allow their installation and use in such a location, Fig. 3.5.4(c).

4. Check capacity against required working radius of the derrick and the precast member's weight, both at the available unloading area and at the point of installation.

Boom rotation can be a full 360 deg. or it is limited by the guy or stiffleg design. As an example, a Chicago boom will rotate 180 deg. when installed on the face of a building and 270 deg. if corner mounted. Some stiffleg derricks are limited by the manufacturers' design to 90- to 100-deg. rotation, which greatly reduces the areas reached with the hoisting apparatus.

3.6 Miscellaneous Specialized Equipment

Various innovations have been devised by precast concrete erectors to accommodate particular site problems regarding access or allowable floor loading. While some types of specialized equipment have become standard, ingenuity often provides economical solutions to difficult or unusual solutions.

Among the many types of specialized hoisting equipment or methods used by precast concrete erectors to accomodate heavy precast units or lack of good access are the following:

1. A specially engineered and fabricated traveling and hoisting apparatus designed to transport and hoist precast concrete units. Better load distribution is often achieved by increasing the number of axles and wheels, thus allowing the vehicle to travel on previously erected floors, see Fig. 3.6(a).

2. A specially designed roof-top crane on a circular track has been used to install precast panels.

3. Helicopters can be used for erection of precast units of small to moderate size, especially where point of placement is difficult to reach. Careful planning is required as the hover time is usually limited. **The safety precautions for helicopter operation shall be rigidly enforced both for helicopter safety and safety of the workmen.** Communication between pilot, ground foreman and top foreman is essential, using both radio and visual signs.

4. A balanced-beam horizontal positioning system (flying jib) which mounts between a crane hook and the load has been used to place panels. This method can be used when limited space or overhanging roof prevent conventional crane setting of panels, Fig. 3.6(b).

5. To erect interior precast units or move them to unusual locations, ingenious methods have been used. Among the most common are the use of telescoping boom cranes with short booms, forklifts, [Fig. 3.6(c)], hand dollies, jacks, chain-falls, machin-

(a) Custom-built carrier fitted with scissor lifts and roller-lined turrets allows the erection of precast elements in areas that are inaccessible to cranes.

(b) Cantilever counterweighted positioning device.

(c) Modified forklift to erect in area with limited headroom.

Fig. 3.6. Specialized lifting equipment.

ery dollies, rollers, greased steel plates, sliding teflon bearings, air supports, etc. As a precaution, when a beam or girder is to be moved by support at its base (e.g., by rolling), sufficient lateral support should be provided to prevent tipping.

Where the size of the project will support the expenditure, standard hoisting equipment can be modified to suit the site condition and/or the allowable floor capacity. Specially designed motorized hauling equipment such as engine-powered, rubber-tired A-frame carriers have been built to facilitate handling and speed the erection process. As a rule of thumb, hand-rigged panels should be kept to a relatively small size and weight if the erection is to be performed economically.

3.7 Operating Equipment

The precast industry is as diversified as the construction industry it serves. Common materials, techniques and equipment can be used with different product types. Hollow-core slabs usually require equipment for barring, shoring, grouting and sometimes welding. Architectural precast usually requires bolted and/or welded connections. Structural precast often requires a combination of each including numerous hand tools such as come-a-longs and jacks in addition to torches and welding equipment. It would be impossible or impractical to list all needs of any given project, but you will find below a general discussion of many of the tools and equipment an erector will likely need.

3.7.1 Welding Machines

All arc-welding processes require a continuous supply of electrical current in sufficient amperage and of proper voltage to maintain an arc. This current may be either alternating (AC) or direct (DC), but it should be supplied to the welding electrode through a device that enables its precise control. Only when the welding current is carefully controlled, can the desired welding arc characteristics including maximum welding efficiency be obtained. The controlling device is called a source or welder. Current may be supplied to it from jobsite power, or developed within it by generators or alternators driven by close-coupled gasoline or diesel engines.

Various types of power sources provide a range of voltage across the welding arc from 17 volts — the minimum voltage for starting an arc — to approximately 45 volts. The current supplied through the power source may range from less than 10 amps to 1,500 amps or more, the higher currents for automatic welding.

For efficient welding, the power source should permit control of arc characteristics needed for a specific job. In one job, a forceful, deeply penetrating arc may be required, while, in another, a soft, less-penetrating arc may be necessary to avoid burn-through. Electrodes are designed for various welding positions and they help compensate for power sources that have no characteristic adjustment.

1. Gasoline or Diesel AC Power Generators/Welders with air-cooled engines are light, compact, portable and can be used as a welder or a generator. **A gasoline or diesel-powered welder shall be used in well ventilated areas due to exhaust fumes.** Consideration should be given to fuel costs, engine noise, maintenance and standby labor costs in certain situations.

   DC air-cooled Engine Welders are medium duty machines which are light, compact and completely portable. They produce AC auxiliary power for lights and hand tools. Step type current control and voltage rheostat are used for precise arc control.

   Heavy duty gasoline and diesel DC engine welders are the time-tested standard for the industry. Control of both current and voltage permits the operator to select arc characteristics needed for the job. AC/ DC auxiliary power is available for lights and hand tools. Low engine speeds and reserve horsepower provide long life with minimum maintenance. Electric start is standard. Consideration should be given to fuel costs, engine noise, exhaust fumes, maintenance and to standby labor costs in certain situations.

   Gasoline or Diesel Machines

   (a) advantages
      - mobility
      - versatility
      - can be used as a generator

   (b) disadvantages
      - leaks oil and fuel
      - exhaust fumes
      - fuel cost
      - maintenance
      - noise

2. Electric machines, constant power inverters, AC and DC rectifiers and transformers, feature crank type continuous current control, good arc stability and soft arc. DC machines have silicon rectifiers with a polarity switch. Electric machines are quiet and require minimal maintenance. There should be a power source at the construction site which may present a hook-up problem.

   Electric Machines

   (a) advantages
      - no noise
      - no mess
      - low maintenance
      - no fuel

   (b) disadvantages
      - requires power source
      - limited portability
      - may require large cable

3. Wire feed machines — Innershield is a semiautomatic process using self-shielded, flux-cored continuous electrodes. In effect, an innershield electrode is a stick electrode turned inside out and made into a continuous wire. All shielding, slagging and deoxidizing materials are in the core of the tubular wire. No external shielding gas or flux is required. Equipment required for this process is a constant voltage DC power source, constant speed wire feeder, gun and cable assembly.

   Wire Feed Machines

   (a) advantages
      - can place considerable weld in single location
      - can make larger fillet welds in a single pass
      - power source can be from any generator
      - no wasted material
      - no special requirements for storage of material

   (b) disadvantages
      - limited portability from power source
      - must maintain the tips
      - requires more preheating; caution: very sensitive to preheat conditions.
      - requires a separate welder certificate.

3.7.2 Torches

Cutting torches may be used with oxygen and three types of preheating gasses.

Acetylene reaches the highest flame temperature, thus making a faster cut. Acetylene produces more slag than propane or natural gas and is more expensive.

Propane gas cutting is used with oxygen in a number of places because of its availability, lower cost and ease of use. **A cutting tip designed for use with propane shall be used.** Do not use an acetylene tip. Acetylene regulators can be used with propane.

If natural gas is readily available in an industrial area, it is a very useful fuel for the preheating flames of cutting torches. Its characteristics as a fuel gas with oxygen cutting operations are much like those of propane. The same cutting tips are generally used for propane and natural gas.

Acetylene

(a) advantages
- quicker cut

(b) disadvantages
- produces more slag
- more expensive

Propane/Natural Gas

(a) advantages
- cheaper fuel
- clean burn

(b) disadvantages
- slower to heat up

3.7.3 Concrete Saws

In the course of erecting precast units, the erector may be required to correct a product fit problem by cutting one or both of the conflicting units. If the correction is minor and is required to continue the erection sequence, it is not unusual that this cutting will be performed by the erector using a cutoff saw. Certain types of product, such as hollow-core slabs, may require more saw cutting particularly for irregularly shaped buildings. **Hand held cutoff saws can be very dangerous and shall be operated by a person trained in their safe use. Ensure that the operator has sufficient safety protective gear.** The manufacturer's operator's manuals usually have pertinent operating and safety information. Also refer to p. 49 in the PCI Erection Safety Manual[17] for other considerations when operating a cutoff saw.

There are several types of cutoff saws. Listed below are some of the advantages and disadvantages of each.

CUTOFF SAWS — Types by power source

1. Gasoline

(a) advantages
- mobility

(b) disadvantages
- high maintenance
- weight
- noise
- exhaust fumes

2. Electric

(a) advantages
- low noise
- no exhaust
- lightweight

(b) disadvantages
- requires power source
- limited power
- limited mobility

3. Hydraulic

(a) advantages
- low maintenance
- faster cut
- considerable power

(b) disadvantages
- expensive
- requires specialized power source

The choice of what type of blade to use depends on the job conditions and the amount of cutting to be done. Dry diamond blades are used most often as long as a minimal amount of steel is to be cut. Proper use and care of blades is not only a safety consideration, but can extend the useful life of the blade. Blade manufacturers have information available. Listed below are the most common blade types with the advantages and disadvantages of each.

BLADES

1. Dry Diamond Blade

(a) advantages
- speed of cut
- constant depth of cut

(b) disadvantages
- dusty
- shorter blade life than wet cutting
- more expensive than wet for frequent cutting

2. Wet Diamond Blade

(a) advantages
- longer blade life
- speed of cut
- no dust

(b) disadvantages
- requires water source
- messy
- slurry may impact other trades

3. Carbide Blade

(a) advantages
- blades are inexpensive
- choice of masonry or steel blades

(b) disadvantages
- dusty
- short blade life
- inconsistent depth of cut

3.7.4 Suggested Tool List

As stated above, the tools required to erect hollow-core slabs are quite different than those an erector will need to erect a parking garage, but there is some commonality even in these extremes. The Precast Erection Truck Equipment and Tool List provided below is meant to be a guide. Each job has its own special needs, particularly with regard to rigging and personnel access equipment. This list does not address tools and equipment required for grouting or drypacking. Conditions and techniques vary so widely that each job should be considered separately.

Precast Erection Truck Equipment and Tool List

LAYOUT AND SURVEY EQUIPMENT

- Transit(s) and Tri-Pod(s) (two are necessary to plumb columns and wall panels)
- Sight Level and Tri-Pod
- Level Rod
- 2' and 4' Levels
- 6' Jamb Level
- 25' Tape
- 100' Tape
- 200' Engineer's Chain
- House Broom and Street Broom
- Chalk Box and Chalk
- Lumber Crayons
- Dry Line
- Clear Spray Lacquer

POWER TOOLS

- Gas-Powered Saw and Blades
- Heavy-Duty Rotary Hammer and Bits
- Medium-Duty Rotary Hammer and Bits
- 7" to 9" Offset Grinder
- (4) 1/2" Offset Grinders
- 1/2" Electric Drive Impact Wrench
- 3/4" Electric Drive Impact Wrench
- Electric Cords — #12-3 Wire
- Ground Fault Circuit Interrupters
- Voltage and Continuity Tester
- Shop Vacuum
- Reciprocating Saw

EQUIPMENT

- 3/4", 7/8", 1", 1-1/8", 1-1/4" Thread Chasers
- Deep-Sockets for Thread Chasers
- Open-End and Box-End Wrenches
- Slugger Wrenches (knocker wrench)
- 16" Pipe Wrench
- 12" and 15" Adjustable Wrenches
- Lineman Pliers
- C Clamps
- Pry Bars — 60" and 66"
- Alignment Bars — 30"
- Miscellaneous Hand Tools and Screw Drivers
- Spoon Bars
- 2 lb., 4 lb. and 6 lb. Hammers
- Rock Pick Hammer
- (2) 1-1/2 Ton and (2) 3 Ton Ratchet Hoists
- (2) 3 Ton Chain Come alongs
- (2) 1/2 and 3/4 Cable Grips
- Tag Lines
- Stepladders
- Assortment of Straight and Extension Ladders
- Welding Machine(s)
- Cutting Torch Gauges, Hose, and Extra Tips
- Tip Cleaner
- Torch Cleaner
- Hose Repair Kit
- Welding Hood(s)
- Burning Glasses
- Electrical Tape
- Welding Gloves
- Welding Sleeves
- Slag Hammers
- Fuel Cans
- Drinking Water Containers
- Shovels
- Wedges — Wood and Steel
- Hydraulic Jack
- Porta-Power Jack with Duck Bill
- Bottle Jacks

SUPPLIES

- Welding Rods
- Saw Blades
- Expansion Anchors
- Tie Wire
- #9 Wire
- Drinking Cups
- Duct Tape
- Welding Lenses
- Spare Electrode Holders (Stingers)
- Oxygen and Acetylene Bottles

SAFETY EQUIPMENT

- First Aid Kit
- Fire Extinguisher(s)
- MSDS and HAZ COM Book
- Body Harnesses
- Lanyards
- Rope Grabs
- Life Line Ropes
- Safety Glasses/Face Shields
- Barrier Tape
- Traffic Cones
- Hard Hats
- Dust Masks
- Spare Safety Latches

RIGGING (Basic)

- (2) 30' x 1' Eye and Eye Slings
- 2 Pair — 3/4" x 20' Thimble eye on Rings
- 2 Pair — 7-1/2 Ton Round Eye Hooks
- (2) 20 Ton Shackles

(8) 8 Ton Shackles
1 Pair — Nylon Slings — 4" x 20' x 2 Ply
Miscellaneous Hooks, Shackles and Chokers
Swivel Lift Plates — Assorted Sizes
Coil Bolts — Assorted Sizes
Burke Lifts — Assorted Sizes
Swift Lifts — Assorted Sizes
(4) 20 Ton Rolling Blocks
(4) 8 Ton Rolling Blocks

3.8 Personnel Access Equipment

3.8.1 General

Construction projects are often large and complicated and access is sometimes difficult. **The selection of access methods shall be determined by safety concerns, the physical constraints of the site, equipment availability and economics.** While engineers should try to locate and design connections where they can be made conveniently from a level surface, there will always be times when personnel must gain access by ladder, mobile elevated platforms or even by crane. It is strongly recommended that personnel access be considered in the conception stages of product and connection development.

This section considers most of the standard methods used in the industry to place personnel at elevated locations under safe working conditions. This section should be used in conjunction with the applicable chapters in the PCI Erection Safety Manual.[17]

3.8.2 Portable Ladders

Ladders are one of the most widely used methods to access locations above the normal working surface. As long as the task is relatively simple, not requiring heavy pulling, pushing, lifting or prying and does not require many tools and materials, ladders are fairly safe and efficient. Many accidents have resulted from trying to do too much from a ladder and/or not setting it up correctly. Refer to the PCI Erection Safety Manual[17] for further considerations when using ladders.

3.8.3 Scaffolds

3.8.3.1 General.

1. **Scaffolds shall be designed for their intended use by a qualified person and erected, moved and dismantled under the supervision of a competent person in accordance with applicable manufacturer's recommendations and 29CFR Part 1926, Subpart L — Scaffolds.** The duties of the competent person may be shared between several individuals. **Each competent person shall possess the qualifications related to his or her area of responsibility, and each shall have the ability and authority to take corrective action.**
2. **Safety instructions shall be included with all scaffolding whether it is owned or leased.**
3. **Scaffolds shall be inspected for visible defects prior to each day's use by a competent person. Such items as solid foundation, stable condition, complete working/rest platforms, suitable anchorage points, required guardrails, loose connections, tie-off points, damaged components, proper access, use of fall protection equipment shall be inspected. Deficiencies shall be corrected before use.**
4. **Do not intermix various manufactured scaffold components unless approved by a competent person.**
5. **Ladders or makeshift devices shall not be used on top of scaffold platforms to increase the height at which employees work.**
6. **Work is prohibited on scaffolds covered with snow, ice or other slippery material except as necessary for removal of such material.**
7. **During high winds, work on or from scaffolds is prohibited unless a competent person has determined that it is safe and that employees on the scaffold are protected by a personal fall arrest system or wind screens. Wind screens shall not be used unless the scaffold is designed for them and the scaffold is secured against wind loads imposed on it.**
8. **Scaffolds shall not be erected, used, or moved closer to insulated energized power lines than the clearance distances required for crane booms. Clearance between the power lines and scaffold shall be not less than 10 ft (3 m).**
9. **Only scaffold grade wood planking, or fabricated planking and decking meeting scaffold use requirements shall be used.**
10. **Platforms at all working levels shall be fully planked or decked with platform units between the front uprights and the guardrail supports.** Space between adjacent units and between the platform and uprights should be no more than 1 in.
11. The front edge of all platforms should be positioned not more than 14 in. (35.6 cm) from the face of the structure being worked on, unless fall protection is provided.
12. **Scaffold platform units, unless cleated or otherwise restrained by hooks or equivalent means at both ends, shall extend over the centerline of the end supports no less than 6 in. (150 mm) and not more than 18 in. (45.7 cm).** A unit may extend more than 18 in. (45.7 cm) over the end support when the unit is designed and installed to support

employees on the extended area without tipping, or guarded to prevent access to the cantilevered ends.

13. **The loads required for the job shall be communicated to the scaffolding company, and the scaffolding company shall communicate the structural limits of the scaffold to the erector. Scaffolds and their components, except for suspension ropes and guardrail systems, shall be capable of supporting without failure, its own weight and at least four (4) times the maximum intended load applied or transmitted to it.**

14. **All frame-to-frame joints shall be pinned.**

15. **Each worker on a scaffold more than 10 ft. (3 m) above a lower level shall be protected from falling.**

16. **Checklist for Scaffolds — General:**
    a. **Safety instructions available.**
    b. **Inspected daily for visible defects.**
    c. **No intermixing of components.**
    d. **Proper clearance from power lines.**
    e. **Proper planking and decking with toeboards.**
    f. **Platforms fully planked or decked.**

The erector shall have each employee who performs work while on a scaffold trained by a person qualified to recognize the hazards associated with the type of scaffold being used and to understand the procedures to control or minimize those hazards. The training shall include the following areas, as applicable:

1. **The nature of any electrical hazards, fall hazards and falling object hazards in the work area.**
2. **The correct procedures for dealing with electrical hazards and for erecting, maintaining, and disassembling the fall protection systems and falling object protection systems being used.**
3. **The proper use of the scaffold, and the proper handling of materials on the scaffold.**
4. **The maximum intended load and the load-carrying capacities of the scaffolds used.**

The erector shall have each employee who is involved in erecting, disassembling, moving, operating, repairing, maintaining, or inspecting a scaffold trained by a competent person to recognize any hazards associated with the work. The training shall include the following topics, as applicable:

1. **The nature of scaffold hazards.**
2. **The correct procedures for erecting, disassembling, moving, operating, repairing, inspecting, and maintaining the type of scaffold in question.**
3. **The design criteria, maximum intended load-carrying capacity and intended use of the scaffold.**

When the erector has reason to believe that an employee has lost the skill or understanding needed for safe work involving the erection, use or dismantling of scaffolds, the erector shall retrain each such employee so that the needed proficiency is regained. Retraining is required in at least the following situations:

1. Where changes at the worksite present a hazard about which an employee has not been previously trained.
2. Where changes in the types of scaffolds, fall protection, falling object protection, or other equipment present a hazard about which an employee has not been previously trained; or
3. Where inadequacies in an affected employee's work involving scaffolds indicate that the employee has not retained the requisite proficiency.

3.8.3.2 Swinging (Suspended) Scaffolds (single and two-point adjustable suspension)

1. **Persons using swinging (suspended) scaffolds shall be instructed in their operation, use, inspection and control.**
2. **On swing stages designed for a working load of 500 lbs. (227 kg), no more than two workers shall be permitted to work at one time. Swing stages designed for 750 lbs. (340 kg) can accommodate up to three (3) workers at one time.**
3. **Suspension scaffolds shall be suspended by wire, synthetic or fiber ropes capable of supporting 6 times the maximum intended load applied or transmitted to that rope with the scaffold operating at either the rated load of the hoist or 2 (minimum) times the stall load of the hoist. All other components, including outrigger beams and tiebacks shall be capable of resisting at least 4 times the tipping moment imposed by the scaffold operating at the rated load of the hoist or 1.5 (minimum) times the tipping moment imposed by the scaffold operating at the stall load of the hoist, whichever is greater.**
4. **Outrigger beams or hooks shall be tied back with a rope equivalent in strength to the hoisting rope installed at right angle to the face of the building or opposing angle tiebacks installed and, whenever possible, secured to a structurally sound portion of the building.**
5. **The supporting rope for single-point adjustable suspension scaffolds shall be vertical between the scaffold and the suspension device, Fig. 3.8.3.2.**

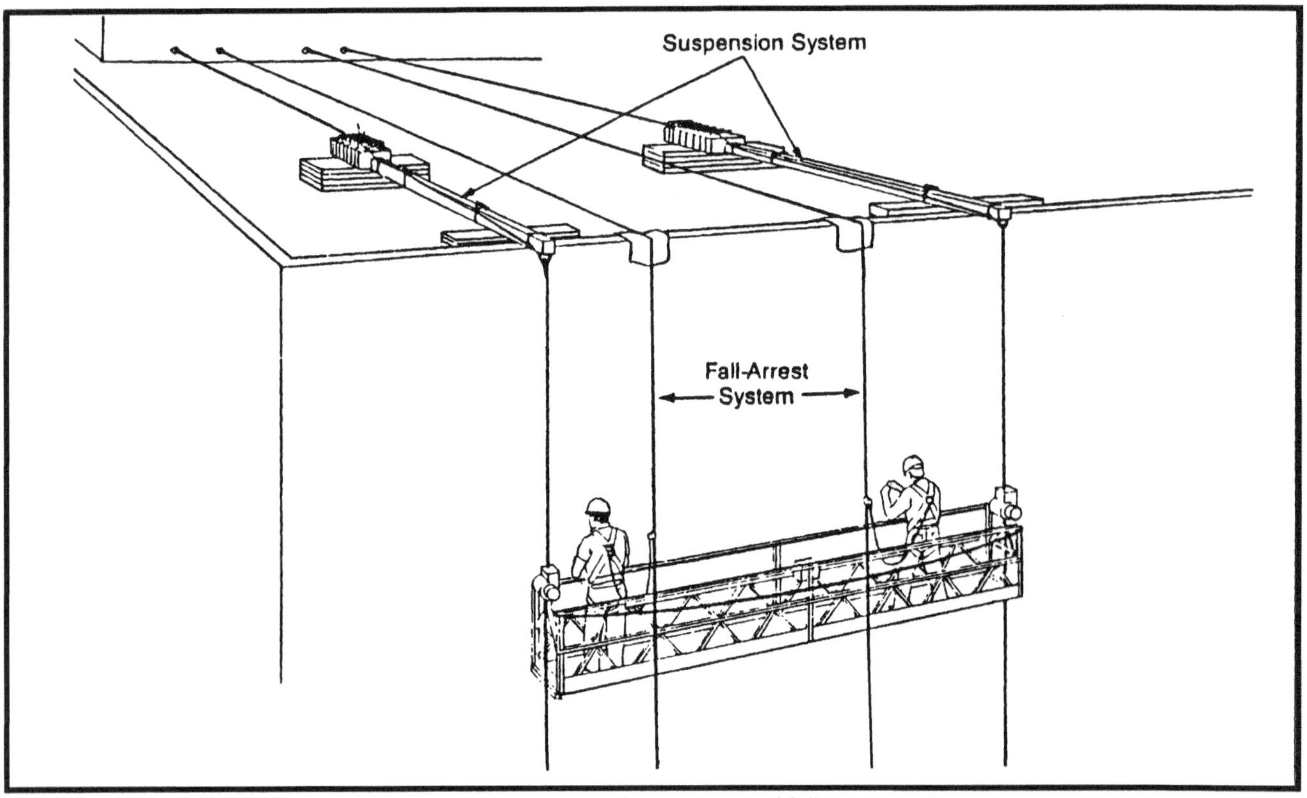

Fig. 3.8.3.2 Two independent means of support — suspension system and fall-arrest system.

6. **Two-point adjustable suspension scaffolds shall be secured to prevent them from swaying based on an evaluation by a competent person.**

7. **Ropes and rope guide assembly shall be inspected and serviced by a competent person prior to each workshift. The use of repaired wire rope and defective suspension ropes shall not be used. The guides shall be replaced when the wear could damage the suspension line.**

8. **Cable clips shall not be used on suspension wire rope lines. Loops in suspension lines shall have thimbles and be either spliced or secured with a mechanically swaged fitting under the supervision of a qualified person.**

9. **Extreme care shall be exercised when working with flame cutting equipment or welding in the vicinity of falls, whether of rope fiber or steel cable. Suspension ropes shall be shielded from heat-producing processes or corrosive substances. All welding leads and electrodes shall be kept away from wire rope and all burned rope shall be replaced immediately. A grounding conductor shall be connected from the scaffold to the structure.**

10. **Workers shall be protected by Type I or Type II guardrail systems. A body belt/harness system shall be worn at all times with Type II guardrail system and properly tied off to an individual lifeline with 1/2 in. (13 mm) nylon lanyards with a 5000 lb. (2268 kg) breaking strength. Lanyard shall be fitted with locking snap hooks or be spliced to rope grabs. Consideration shall be given to using lanyards with built-in shock absorbers to help reduce the impact of a fall. The employee shall be tied off to the vertical life line at all times during ingress and egress of platform.**

11. **Mechanically powered hoisting devices shall be equipped with a positive device to prevent staging from falling freely.**

12. **Consideration shall be given for communication and rescue in case of a system failure.**

13. **Each individual vertical lifeline shall be suspended securely and independently from staging and extend from the anchorage point to the ground or a safe landing place above the ground.** They should hang as close to vertical as possible. **Freefall safety devices shall not be overridden.**

14. **Prior to use each day, a competent person shall thoroughly inspect the staging including outrigger beams, counter weights, lifelines and mechanical devices.**

15. **Support cables shall extend from the staging to the ground or be provided with a positive stop to prevent staging from running off the end of the supporting cables.**

16. **Staging shall be securely anchored to the structure when not in use.**

17. **Local regulations shall be verified regarding use of swing stages. Some areas may require special riggers license.**

18. **Checklist for Swinging (Suspended) Scaffolding:**

    a. **Required components properly rigged-up and in good condition.**

    b. **Operator knowledgeable and competent to operate the equipment.**

    c. **Fail-safe devices such as rope grabs, secondary safety devices and overspeed controls installed and operating.**

    d. **Numbers of workers on the staging shall not exceed the manufacturer's design or recommendations.**

    e. **Guardrails and toeboards installed.**

    f. **Body belt/harness system tied to lifeline.**

    g. **Power supplies for climbers adequate, grounded and secured.**

    h. **All tiebacks for outrigger beams, parapet clamps and lifelines properly secured to adequate anchorage.**

    i. **Adequate number of counterweights securely attached to outrigger beams.**

    j. **Fiber ropes protected from chafing and abrasion.**

    k. **Emergency rescue arrangements planned, prepared and communicated to everyone involved.**

    l. **Access to and from the work area planned and arranged.**

    m. **Wind conditions evaluated.**

    n. **Properly rigged, tied and anchored to prevent swaying.**

3.8.3.3 Supported Scaffolds.

1. **Scaffolding shall be erected on sound, level, rigid footing capable of carrying the maximum intended load without settling or displacement. Scaffolds erected on any type of soil shall be on a base plate and mudsill.** The mudsill should be a minimum of 2 in. x 10 in. (50 mm x 25.4 cm) plank (full size) and should be continuous under at least two consecutive supports. **Scaffold feet shall rest centrally on the mudsill and the sill should project at least 2 ft (0.6 m) beyond the scaffold foot at ends or where individual sills butt together.**

2. **All uprights shall be assembled plumb and securely braced (cross and/or diagonal braced) as per manufacturer's recommendations.**

3. **All fittings and gear shall be properly installed including base plates or wheels.**

4. **When scaffolding height is more than 4 times its width, scaffolding shall be securely braced against the structure to prevent any lateral movement.** Restraints are recommended at least every 20 ft (6.1 m) in height and at least every 30 ft (9.1 m) in width and at the ends of the scaffold runs. **The tie-in systems shall be designed to resist the loads that are likely to be applied.** These may consist of wind loads or dynamic loads caused by work being done on the scaffold. **Such loads are difficult to estimate but at least 15 lbs/ft (22.3 kg/m) of platform length at each working level shall be allowed for scaffolds carrying workers, their tools and light materials.** In addition, a safety factor of 2 is usually applied to the design of the tie-in system.

5. **Scaffolding shall be plumb, level and square. The scaffold frame shall be checked after each tie is added.**

6. **Fall protection shall be provided for workers erecting or dismantling supported scaffolds where a competent person has determined the installation and use is feasible and does not create a greater hazard.**

7. **Employees shall use body belt/harness systems or Type I guardrail systems. Body belt/harnesses shall be attached by a lanyard to a lifeline or scaffold structural member.**

8. **Guardrails between 38 and 45 in. (96.5 and 114.3 cm) high (where the guardrail is the primary means of fall protection) shall consist of a top rail, intermediate rail and toeboard (when required).** Top rail may be 36 in. (91.4 cm) high where personal fall arrest systems are the principal means of fall protection and guardrail systems serve primarily to mark the perimeters.

9. **An access ladder with rails projecting 3 ft (0.9 m) above the platform of the scaffold or equivalent safe access shall be provided. Climbing of the scaffold bracing shall be prohibited.**

10. **Checklist for Supported Scaffolding:**

    a. **Ensure that required components are available in good condition and sufficient quantity:**

    (1) Mudsills for scaffolds erected on soil.

    (2) Base plates and screw jacks.

    (3) Vertical and horizontal braces.

(4) Platform material to fully deck-in the working level.

(5) Frame connecting devises such as toggle pins (banana clips) and pig tails.

(6) Guardrails (including toeboards).

(7) Ladders, whether built-in or portable.

b. Scaffolds shall be erected and dismantled under the supervision of personnel knowledgeable and experienced in this operation (competent person).

c. Scaffolds components shall be inspected for damage or corrosion before installation and prior to each day's use.

d. All required components shall be installed as scaffolds are erected.

e. Scaffolds shall be erected according to manufacturer's recommendations.

f. Scaffolds shall be adequately braced in the horizontal and vertical planes.

g. Working platform shall be completely decked.

h. Ensure that lumber scaffold planks are of scaffold grade and at least 2 in. x 10 in. (50 mm x 25.4 cm) (nominal) in cross-sections and free of noticeable defects such as loose knots, splits or rot.

i. Frame of scaffold shall be tied into the building or structure at a maximum of 20 ft (6.1 m) vertically and 30 ft (9.1 m) horizontally.

j. Frame shall be pinned together on towers over two frames high.

k. Scaffold planks shall be secured to the scaffold or cleated to prevent them from sliding.

l. Planks shall overhang supports by at least 6 in. (150 mm) but not more than 18 in. (45.7 cm).

m. Platforms shall be secured to the scaffold to prevent sliding and wind uplift.

n. Working platform shall be protected by proper guardrails, midrails and toeboards.

o. Body belt/harness system shall be attached by lanyard to lifeline or scaffold structural member.

p. Scaffolds shall be erected reasonably level, plumb and square.

q. Scaffolds over 125 ft (38.1 m) high shall be designed by a registered professional engineer and the scaffold erected according to the design with the drawings kept on site.

r. Scaffolds shall be clear of overhead electrical wires by a minimum of 10 ft (3 m).

s. Portable ladders to scaffolds shall be properly tied off and positioned to not tip the scaffold and the rails shall extend at least 3 ft (0.9 m) above the working platform.

t. Scaffolds shall be free of slipping hazards such as spills, oil, grease, ice and snow.

u. Scaffolds shall be free of tripping hazards such as tools, equipment, litter, material or personal belongings.

v. Scaffold planks shall not be painted.

3.8.3.4 Mobile (Rolling) Scaffolds.

1. Mobile scaffolds being used in a stationary manner shall be secured against unintentional movement.

2. When employees are working on mobile scaffolding, the casters shall be locked.

3. Caster and wheel stems shall be pinned or otherwise secured in scaffold legs.

4. The force used to move a mobile scaffold shall be applied as close to the base as possible, but not more than 5 ft (1.5 m) above the supporting surface. Surfaces over which the scaffold is to pass shall be free of obstructions and openings that may cause the scaffold to tip.

5. All uprights shall be assembled and securely braced (cross, horizontal, or diagonal braces) per manufacturer's recommendations.

6. Employees shall not be allowed to ride on scaffolds unless the following conditions are met:

   a. The surface over which the scaffold will pass shall be within three degrees of level, and free of pits, holes and obstructions.

   b. The maximum height to base width ratio of the scaffold during movement shall be two to one or less. Outrigger frames may be included as part of the base width dimension.

   c. Outrigger frames, when used, shall be installed on opposite sides of the scaffold.

   d. Tools and materials shall be removed from the platform unit, or secured to prevent movements or toeboards shall be installed on all sides of the scaffold.

   e. Employees shall not be on any part of the scaffold which extends outward beyond the wheels, casters, or other supports unless the scaffold was designed for this use.

f. Employees on the scaffold shall have advance knowledge of the movement.

7. Checklist for Mobile Scaffolds:

   In addition to checklist for supported scaffolding the following shall apply:

   a. Wheels locked before work begins.
   b. Casters securely pinned to legs.
   c. Outriggers installed, if required.

3.8.4 Mobile Elevating Work Platforms (MEWP) — Aerial Lifts or Manlifts

Whether needed to weld clips at the exterior of a building or to connect the underside of a difficult-to-reach spandrel beam, mobile elevating work platforms (MEWP) have become necessary tools for the precast erection industry. Aerial work platform usage is job specific; each application requires a unit with particular capabilities and size With work heights now reaching 150 ft (45.7 m), and with both rough terrain and electric power capabilities, these versatile machines allow personnel to easily and safely position themselves for work at heights. For safer operation, units should be equipped with slope level interlocks that sound a warning or lock-out the lift function if the unit reaches a predetermined slope.

1. An operator shall be trained in the safe use and basic limitations of the specific MEWP before being allowed to use them.

2. Operators shall be instructed in the daily maintenance and inspections for which they are to be responsible.

3. When MEWP is to be used, ensure that adequate access and working areas are provided.

4. Each unit shall be equipped to prevent unintended motion (brakes shall be set) when in use. (When outriggers are used, they shall be positioned on pads or a solid surface).

5. The maximum platform height of units that only elevate in the vertical plane, without any articulation, shall not exceed four times the minimum base dimensions unless equivalent stability is provided. When greater heights are necessary, properly fitted outrigger frames shall be provided.

6. An out-of-level sensor shall be used to disable boom functions when the unit is tilted at an unsafe angle, usually 5 degrees.

7. Toeboards shall be provided on all sides of the platform except across access openings.

8. The elevating assembly shall be equipped and maintained so that it will not allow a free descent or an uncontrollable fall in the event of the assembly's failure. Any unit equipped with a powered elevating assembly shall be supplied with a clearly marked means for emergency lowering that is accessible from the ground level.

9. Outriggers and stabilizers shall be constructed to prevent unintentional retraction and result in an unstable work platform.

10. Assure before and during movement of units that:

    a. The platform has been lowered to base level and outriggers are in stowed position.
    b. Tools and materials on the platform have been secured from falling or have been removed.
    c. The area the unit is being moved through has a firm footing and is cleared of obstructions.

11. The area surrounding the unit shall be cleared of personnel and equipment before the platform is lowered to prevent personnel from being under the descending units, and from being struck or injured by the platform itself or by other pieces of the equipment.

12. Operator shall test the machine before use — first from the ground then from the platform at a low height to be sure it is operating properly.

13. Manufacturer's recommended platform load capacity shall not be exceeded. Load capacity shall take into account the combined weight of all personnel, tools, fixtures, and accessories that are on the platform.

14. The load shall always be distributed evenly over the platform floor area.

15. Workers on platform shall wear a body belt/harness attached to the basket or boom with a lanyard.

16. Ensure workers always stand with both feet on the floor of the platform and never lean over guardrails to perform work.

17. Work platforms shall not be operated within 10 ft (3 m) of energized high-voltage power lines.

18. Platform gate shall be secured immediately after entering.

19. The boom shall be elevated only when the work platform is on a firm, level surface.

20. **When the MEWP is used for egress to an elevated surface, the employee shall be tied off when making the transfer.**

21. **Checklist for mobile elevating work platforms:**
    a. **Operators trained in uses and limitations.**
    b. **Maintained and inspected daily.**
    c. **Secured to prevent unintended movement.**
    d. **Out-of-level sensor.**
    e. **Toeboards.**
    f. **Equipped to prevent free descent.**
    g. **Means for emergency lowering.**
    h. **Body belt/harness system worn and attached to platform or boom.**
    i. **Area cleared of workers before platform lowered.**

The following is a description of the different types of mobile elevating work platforms available and the particular attributes of each machine.

3.8.4.1 Telescoping Boom Platforms (Fig. 3.8.4.1). The American National Standards Institute (ANSI) defines telescoping boom machines as "self-propelled integral chassis aerial platforms having a platform that can be positioned completely beyond the base and used to position personnel, along with their necessary tools and materials, at work locations." (ANSI A92.5).[22] Telescoping boom platforms come in two basic designs: straight and articulating boom models. Straight boom machines are similar to small hydraulic cranes with telescoping booms that are lowered and raised by a hydraulic cylinder. Articulating or knuckle-boom machines have joints, allowing the boom and platform to pivot into numerous positions unobtainable with a straight boom platform. Both types of machines now have booms which can extend to 150 ft (45.7 m). There are also boom-lifts featuring a combination, such as a straight boom combined with an articulated jib that allows them to reach into obstructed or confined areas.

Fig. 3.8.4.1 Telescoping boom platform.

3.8.4.2 Self-Propelled Vertical Aerial Platforms (Fig. 3.8.4.2). Commonly referred to as scissor lifts, self-propelled vertical aerial platforms have gained tremendous acceptance in the construction industry over the last decade. With the introduction of machines small enough to fit in a standard material hoist, yet capable of supporting 750 lbs. (340 kg) and powered with electricity, in certain situations these lifts have made ladders and pipe scaffold equipment obsolete. Because of their ability to quickly and efficiently raise multiple personnel, as well as equipment and materials to elevated work positions, powered vertical aerial platforms are now becoming a widely utilized piece of equipment in the precast erection industry.

Unlike telescoping boom machines, vertical aerial lifts have a platform that cannot be positioned completely beyond the base. However, newer machines have extendible platforms that cantilever out to 5 ft (1.5 m) for extended reach. Work heights for self-propelled vertical aerial platforms range from 12 to 50 ft (3.7 to 15.2 m). Some scissor lifts have a "pothole" protection system to prevent the lift from tipping should one of the wheels drop into a depression.

3.8.4.3 Manually-Propelled Vertical Aerial Platforms (Fig. 3.8.4.3). Unlike self-propelled vertical aerial platforms, these lifts lack motorized power for travel and require personnel to exit the machine before it is maneuvered to another position. Small, light, and compact, such manually-propelled units are sometimes ideal for tight, congested work areas. Elevated either manually or hydraulically, the platform can be raised to heights in excess of 40 ft (12.2 m).

Fig. 3.8.4.2 Self-propelled vertical aerial platform.

Fig. 3.8.4.3 Manually-propelled vertical aerial platform.

DIVISION 4 — SAFETY

4.1 General

A safety program is an important element of every erection operation and is required by OSHA Safety and Health Standards (29 CFR 1926). **The erector shall develop a written safety program.** It should be continually reviewed and updated so that it contains the latest strategies and technologies available to create a safe and healthy work environment at the construction site. **The written safety program shall define the company commitment to workplace safety and provide the structure for achieving a zero accident culture.** The program should be carefully planned to ensure good safety practices, general housekeeping and review of job related accidents and injuries. The applicable provisions of the PCI "Erection Safety Manual for Precast and Prestressed Concrete," MNL-132[17] should be used to develop such a program.

This section is intended to outline the more important safety practices for erectors in a broad manner but cannot be considered to include every conceivable hazard that may be present on a jobsite.[23] However, recognition of hazards, the establishment of good safety practices and the requirement that all personnel abide by safety rules will result in a more efficient and safer operation.

A company's safety beliefs should be simple and straightforward, such as:

- All injuries can and should be prevented.
- Injury prevention applies to all personnel on the jobsites as well as the public.
- Erector safety is consistent with overall company safety policy and is good business.

At the heart of every safety program there shall be strong management commitment. This commitment, expressed in writing from the highest level of the management hierarchy, will provide the environment for the development of a safety culture. Management cannot create this culture without full participation from the work force. An effective program should include:

1. A clearly stated policy which demonstrates the relative importance of safety and health protection in relation to other corporate values such as quality, productivity, and profit. This policy must be emphatically and frequently communicated to every employee in the company.

2. Clearly defined goals for the program with equally clear, quantifiable objectives for reaching them.

3. A strong and visible leadership role by top management in the implementation and maintenance of an effective program. This will underscore the corporate commitment to the development of a safety culture.

4. Clearly defined and assigned responsibilities for managers, supervisors and employees for all aspects of the program. Each must be empowered with the authority to achieve this end. Each must also have the resources necessary to fulfill his or her commitment to the program.

5. Holding all levels of management responsible and accountable to maintain the level of commitment necessary to consistently achieve the highest standards of safety performance. Performance appraisals should be tied to each individual's contribution to the program using quantifiable goals and objectives. This may include attendance at safety meetings and employee training sessions, participation in inspections and investigations, safety-related contacts with supervisors and employees, and precasters.

   Accountability may also involve the assignment of accident costs directly to the project where losses occur. Some erectors prorate all or some insurance premium and uninsured loss costs and charge them to the appropriate profit center. This emphasizes the principle that even a project completed ahead of schedule and under budget may actually reduce corporate profits if all losses due to accidents are applied to the job.

6. Employee involvement in the development of a safety culture. A reliable system should be provided for employees to notify management personnel about conditions that appear hazardous, (including defective or unsafe materials, tools or equipment, unsafe working practices, and co-workers who appear ill or otherwise unfit to work safely). Employees should receive timely and appropriate responses and be

encouraged to use the system without fear of reprisal. This utilizes employee insight and experience in safety and health protection and allows employee concerns to be addressed. When employees believe that management is committed to the safety program and they are given the opportunity to be a part of the process of developing a safety culture, achieving zero accidents will seem realistic.

The safety program shall require that all accidents resulting in injury to workers and/or damage to equipment or product, regardless of their nature be investigated and reported. Near misses shall also be reported and investigated. It is an integral part of every safety program that documentation shall be accomplished immediately so that the cause and means of prevention can be identified to prevent a reoccurrence. In addition, if an accident has resulted from an employee's failure to follow job procedures, this shall be reviewed with all employees performing the job.

The elements described above provide the philosophic framework for the development of an excellent safety program. The written program will reflect these elements and will provide specific procedures to achieve desired results.

Written safety procedures should require weekly inspections of each site. **Prior to the start of work, and during construction, a competent person shall make regular surveys of the site conditions to determine the potential hazards, both physical hazards (such as equipment and housekeeping deficiencies), and personnel hazards (such as unsafe procedures and activities), and the safeguards necessary to erect precast concrete in a safe manner.** This will help ensure that each site is maintained within compliance of applicable rules and procedures. Information gathered during inspections will allow management to communicate hazard information to employees in a timely fashion. **Good housekeeping shall be practiced at all times. All hazardous debris shall be removed from the vicinity of the work area not later than the end of each shift.**

Some hazards can be engineered-out of the erection process. However, many hazards will remain. **Precautions shall be taken to minimize the element of employee exposure.** Where exposures do exist, safe operating procedures can eliminate the likelihood of injury. This may involve the use of administrative controls, such as a Fall Protection Plan, to reduce the duration of exposure and the use of personal protective equipment.

The erector shall understand and comply with applicable health and safety laws, regulations and contract provisions, and provide a qualified work force that is trained to do the job. The erector shall be responsible for supplying tools, equipment and materials needed to do the work safely, and for supervising and controlling the actions of their employees while on-site.

4.2 Employee Training

All supervisors shall be trained to recognize and correct hazards and to identify and respond to unsafe behavior. Training shall focus on prevention through careful planning, and emphasize the importance of the supervisor's role in safety. This will include the regulatory obligation for them to recognize and control jobsite hazards. It will also underscore the significance that the direct and indirect costs can have on their company's financial results. **To prevent accidents, workers shall be trained in proper practices and procedures, rules and regulations, and be motivated to work safely. Employees shall sign and date an acknowledgment sheet verifying that they have received, read and understand the company safety policies. Employee training programs shall be designed to ensure that all employees understand and are aware of the hazards to which they may be exposed and the proper methods for avoiding such hazards.**

Jobsite safety meetings shall be held at the beginning of a project to ensure the workers are familiar with company safety policies, anticipated hazards, safe job procedures and requirements of the upcoming erection work, including the Erection Safety Plan. Indoctrination of new employees and long-term employees who are new to the jobsite is particularly important. **In addition, erection crew safety meetings, such as toolbox meetings, shall be conducted weekly to discuss topics designed to increase worker safety. These meetings shall be documented as to attendance and subject matter and all employees shall be required to attend.**

Upon start of a project, employees shall be made familiar with the following:

1. **Identity of first aid representative.**
2. **Location of jobsite telephones for emergencies.**
3. **Location of emergency phone numbers and addresses of nearby doctors, ambulance services, hospital, police and fire departments, if 911 service is not available.**
4. **First aid equipment.**
5. **Bloodborne pathogens exposure control plan.**
6. **Location and proper use of fire extinguishers.**

As part of the company's safety program, each work shift shall have access to one person on site who is trained and certified to administer first aid and CPR. This may necessitate a meeting of the various supervisors present on a project to reach an agreement concerning the most expeditious means of administering first aid to accident victims. Arrangements can be made in advance for the various contractors to cooperate in providing first aid to anyone who may require it during working hours. It is necessary, however, that such arrangements be established before actual work on the project begins. The first aid kit should be checked at least weekly to ensure that the expended items are replaced.

The erector shall develop a specific written hazard communications program. The written program shall include a chemical inventory list and establish the methods to provide information to employees on the following:

1. **Container labeling.**
2. **Other forms of hazard warning.**
3. **Material safety data sheets (MSDS) — how sheets are to be maintained and how employees obtain access to them.**
4. **The requirements of the hazard communication standard.**
5. **Hazards associated with non-routine tasks.**

A training program shall be established for employees before they work with hazardous chemicals and when new hazardous chemicals are introduced on the jobsite. Employees shall be trained to read and understand the placard product warning signs and other safety warning information.

All personnel shall be required to use the proper tools and equipment, including safety harnesses, for their assigned tasks and be required to wear the necessary proper personal protective clothing and safety hats, and eye, foot, and hearing protection.

All personnel involved with the operations of cutting, sanding or grinding concrete shall wear an appropriate face mask to avoid respiration of fine silica dust.

Exposure to bloodborne pathogens may occur during the administration of first aid to an injured employee or from opening a contaminated first aid kit. **All human blood and body fluids shall be treated as if they are known to be infectious for Hepatitis B Virus (HBV), Human Immunodeficiency Virus (HIV) and other bloodborne pathogens.**

A written bloodborne pathogens exposure control plan shall be developed along with guidelines for cleaning, identifying the method of documentation, location of cleaning supplies and the handling of cleaning wastes. Personal protective equipment (PPE) shall be available and used whenever exposure is present. All employees shall be properly trained before starting work, after program changes, and yearly as a review in the exposure control plan and in the care, use and disposition of the PPE. The training shall be documented and records kept for 3 years. Medical records of employees shall be kept confidential and maintained for the duration of their employment plus 30 years.

4.3 Equipment Safety Programs

Machines, tools, materials, or equipment identified as unsafe shall be tagged or controls locked to make them inoperable or they shall be removed from the jobsite.

4.3.1 Crane Safety and Maintenance

A copy of the maintenance schedule and inspection record shall be on the crane. The crane logs shall be checked to ensure that inspections are being made. Only designated individuals shall signal the crane. If the operator cannot see the signal person, a radio with a dedicated frequency shall be used or a person shall relay the signal to the operator. A load chart shall be on the crane and the crane shall be level and on stable ground with outriggers properly cribbed.

4.3.2 Electrical Equipment

All electrical wire and equipment in use shall be of a type approved by the Underwriters Laboratories, Inc. Installation and maintenance shall comply with pertinent provisions of the National Electrical Code NFPA-70.[24] All electrical circuits and equipment shall be properly grounded in compliance with the Code, requiring a three-wire polarized system on all extension cords and receptacles. Ground fault circuit interrupters or an assured equipment grounding conductor program shall be used. Electrical cords shall be in good condition and not frayed. Electrical cords shall be placed to minimize tripping hazards. Power tools shall be checked before each use for proper operation.

4.3.3 Lifting Devices

The weight of all precast concrete components to be lifted and the type of lifting devices provided in the product shall be obtained, if not shown on erection drawings.

1. **Lifting hardware shall be compatible with the product lifting inserts. The size of the device shall take into consideration increased load component resulting from pulling at an angle.**
2. **Lifting bolts shall be long enough to have the number of threads engaged as recommended by the hardware manufacturer.**

3. **The lifting hardware and rigging shall be able to take the load in the required direction or directions throughout the lift.**

4. **Check to determine if lifting loops or inserts have the capacity to serve as an anchorage for personal fall arrest equipment, when required.**

4.3.4 Rigging

The weight and size of all precast concrete components to be lifted shall be obtained. There will be times when it shall be necessary to obtain the center of gravity of certain pieces.

1. **Spreader cables shall be adequately sized to handle the piece, taking into consideration the angle of the lift, i.e. a cable pulling at an angle does not have the same lifting capacity as the same cable pulling in the vertical direction.**

2. **Rings, shackles and hooks in the lifting setup shall be sized to handle the load and at the same time be compatible with the size limitation of the lifting device and the hook on the crane.** For example, it is not unusual for the ring of a spreader to be too small to fit on the hook of a large crane; in which case, a large shackle or a short pennant cable will be needed with the appropriate hook. On the other hand, it is not unusual for the spreader hooks to be too large to fit the lifting loops or other devices provided, and other rigging hardware will have to be introduced. **Lifting hardware shall be capable of supporting at least five times the maximum intended load applied or transmitted to the lifting hardware.**

By their nature, some pieces will require steel or nylon chokers which may require protection from sharp edges.

It is not unusual to have to lift pieces at more than two locations. In these cases, it is generally a good idea to review the rigging scheme with the precast concrete designer to ensure that the piece will not undergo unsafe loading as it is being lifted. **Consideration shall be given to the direction of the pull on the lifting device throughout the entire process of making the lift to ensure that neither the lifting hardware nor the rigging will be over-stressed.**

When rotating a piece in the air with two crane lines, the center of gravity shifts during the rotation process. Care should be taken to ensure that the lower lifting devices and rigging will handle the entire weight of the piece.

Cables are often used in the basketed form when erecting hollow-core slabs. **When used under these conditions, the erector shall pay particular attention to two things:**

1. **Cable chokers shall be long enough to ensure safe seating in the hooks, i.e., angle not less than 45 deg.**

2. **Spreaders connected to the chokers shall be long enough so that the spreader angle is no less than 60 deg., otherwise the choker could slip.**

An in-depth check of all initial rigging set-ups shall be made and daily visual inspection shall be performed. Rigging inspection reports should be available.

Wire ropes, slings, shackles, and other rigging equipment shall be kept in good condition. All custom-designed lifting equipment shall be marked with safe working load.

Personnel shall remain clear of precast concrete members being lifted into position except those workers who are actively engaged in the connection of the member.

The erector should have a rigging diagram, either as part of the erection drawings or separately attached, for non-typical picks.

4.3.5 Welding Equipment

1. **Welding areas shall be ventilated to protect workers from noxious welding fumes. Welding and/or flame cutting shall not be allowed in the vicinity of combustibles without fire-watch or protection of combustibles. Oxygen/acetylene cylinders shall be stored separated in an upright position.**

2. **Gasoline or diesel powered welding machines shall be operated in a well ventilated area. The gasoline or diesel powered welding machines shall be shut down while fueling.**

3. **Welding leads shall have no breaks in the insulation for the first 10 ft (3 m) from the end holding the electrode. The end attached to the machine shall be free of breaks in the insulation to prevent accidental arcing around the fuel. All breaks in the insulation shall be repaired to prevent accidental arcing.**

4.3.6 Ladders and Scaffolds

Ladders and scaffolds shall meet OSHA requirements. All ladders shall be checked daily for proper use and for defects. An in-depth check shall be made of all initial scaffolding/staging installations and daily checks thereafter.

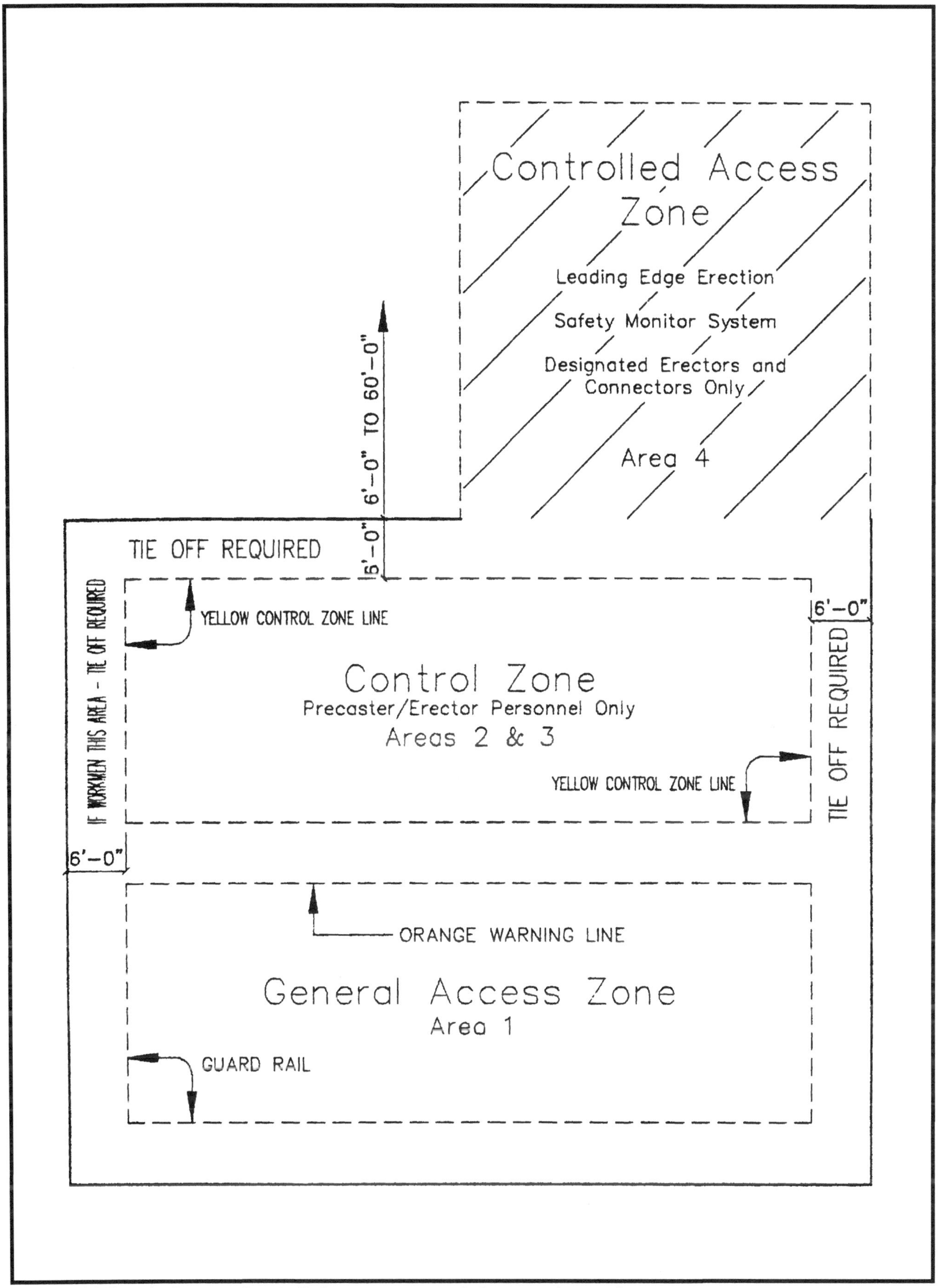

Fig. 4.4.3(a) Sample diagram of controlled access zone, control zone and general access zone.

4.4 Building and Jobsite Safety

The precast concrete erection sequence shall be reviewed with GC/CM for not only precast concrete erection, but for other trades that will be working on the site at the same time. Ensure that cranes and trucks can safely move on to the site without being limited by open excavation, overhead power lines or underground utilities (new or existing), or underground tanks or tunnels. Verify that access from public ways can be accomplished without jeopardizing safety of the public.

Overhead power lines are an extreme hazard to safe crane operation. **Ensure that overhead power lines do not limit the crane throughout its entire operation, including breaking the crane down at the end of the job. Determine if power lines must be de-energized or moved prior to starting erection, refer to the PCI Erection Safety Manual, MNL-132.[17]**

4.4.1 Guying and Bracing

The overall building configuration along with erection drawings and erection safety plans shall be reviewed to ensure that the structure will remain stable during the construction phase. All structures or components that may have a stability problem shall have a plan for the temporary guying or bracing of the structure during erection.

(b-1) Member arriving at leading edge.

(b-3) Erectors positioning member.

(b-2) Erectors taking control of member.

(b-4) Final alignment of member.

Fig. 4.4.3(b) Erection sequence for a double tee. Employees working in controlled access zone using the safety monitoring system.

Guying and bracing required for day-to-day erecting shall be determined by the jobsite competent person.

Verify that the sequence of construction allows for safe swinging of the load from truck to the point of erection, without interfering with guy cables or braces.

4.4.2 Erection Connections

The jobsite competent person shall determine:

1. Unless shown on erection drawings, the minimum number of connections that must be made on a member prior to releasing the crane or hoisting equipment.
2. As part of the consideration for structural stability, how many temporary or permanent connections shall be made during the erection process and what connections may be made as follow-up work.
3. The minimum amount of grouting or drypacking that is required under columns or wall panels prior to erecting each succeeding level.
4. Connections and other critical aspects of construction that can be accomplished by employees working in safe conditions.

4.4.3 Fall Protection

Ensure that the designated erectors have been selected by the competent person. Ensure that all personnel have received or are given fall protection training in accordance with 29 CFR Part 1926, Subpart M — Fall Protection. Ensure that all material and equipment required for protecting employees against a fall is available and used. Figs. 4.4.3(a) to (h) illustrate some of the fall protection procedures used during the erection of precast concrete.

Controlled access lines, perimeter protection, barricades, floor and roof hole covers, shall be checked daily to ensure they are in proper position.

Fig. 4.4.3(c) Erected structure showing the control zone and the general access zone.

4.4.4 Protection of Personnel on Lower Floors

The anticipated overall construction sequence and crew sizes shall be reviewed to ensure that each stage can be safely accomplished with minimum impact on concurrent work of other trades, as well as that of the erecting crew.

The erection sequence shall not have one operation working directly over another nor shall product be swung over other work areas. In multistory structures, follow the guidelines developed for the project to keep a designated number of floors clear of personnel under the work area. It is recommended that personnel are not allowed under the area being erected and under the crane swing radius.

Fig. 4.4.3(d) Control zone lines, hole covers and perimeter protection shall be checked daily to ensure they are in proper position.

Fig. 4.4.3(e) Horizontal lifeline (trolley line) used in conjunction with rope or cable grab to tie off.

Fig. 4.4.3(f) Employee with lanyard tied off to lifting device in precast unit.

Fig. 4.4.3(g) Workers receiving, positioning, and making initial connections of precast concrete cladding shall work behind perimeter protection or be tied off if they are working within six feet of unprotected perimeter.

Fig. 4.4.3(h) Specialized working platform complete with guard rails.

DIVISION 5 — TOLERANCES

5.1 General

The precast concrete manufacturer and erector are responsible for meeting the specific tolerances in the manufacturing and erection processes, respectively.

The architect/engineer should be responsible for coordinating the tolerances for precast concrete work with the requirements of other trades whose work adjoins the precast concrete construction. In all cases, the tolerances must be reasonable, realistic and within generally accepted limits. It is recommended that the designer review proposed tolerances with precasters and erectors prior to deciding on the final project tolerances. The precast manufacturer, GC/CM, and erector must in turn carefully monitor tolerances in order to construct the structure as designed.

Tolerances are divided into three categories:

- product tolerances
- erection tolerances
- interface tolerances

Product, interface and erection tolerances should be taken into account during the design phase of each project to assure satisfactory member installation.

Tolerance is defined as a specified permissible variation from a specified dimension or quantity. A tolerance may be expressed as an additive (+) or subtractive (-) variation from a specified dimension or relation or as an absolute deviation from a specified relation. Tolerances should be established for the following reasons:

1. **Structural.** To ensure that the structural design accounts for factors sensitive to variations in dimensional control. Examples include eccentric loading conditions, bearing areas, locations of reinforcement, hardware and hardware anchorage.
2. **Feasibility.** To ensure acceptable performance of joints and interfacing materials in the finished structure and to ensure that designs and details are dimensionally feasible from production and construction capabilities.
3. **Visual effects.** To ensure that the variations will be controllable and result in an aesthetically acceptable structure.
4. **Economics.** To allow ease and speed of production and erection by having a known degree of accuracy in the dimensions of the precast concrete units.
5. **Legal.** To avoid encroaching on property lines and to establish a tolerance standard against which the work can be compared in the event of a dispute.
6. **Contractual.** To establish a known acceptability range and also to establish responsibility for developing, achieving and maintaining mutually-agreed tolerances.

It should be understood by those involved in the design and construction process that the listed tolerances in this Manual be considered as guidelines for an acceptable range and not limits for rejection. If specified tolerances are met, the units should be acceptable. If these tolerances are exceeded, the unit may still be acceptable if it meets any of the following criteria:

1. Exceeding the tolerance does not affect the structural integrity, architectural performance of the unit or other trades.
2. The unit can be brought within tolerance by structurally and architecturally satisfactory means.
3. The total erected assembly can be modified reasonably to meet all structural and architectural requirements.

When a project involves particular features sensitive to the cumulative effect of generally accepted tolerances on individual portions, the architect/engineer should anticipate and provide for this effect by setting a cumulative tolerance or by providing escape areas (clearances) where accumulated tolerances can be absorbed. One example of how this can be achieved is by limiting columns to 3-story heights. This creates an adjustment area every third floor The consequences of all tolerances permitted on a particular project should be investigated to determine whether a change is necessary in the design or in the tolerances applicable to the project or individual components. For example, there should be no possibility of minus tolerances accumulating so that the bearing length of members is reduced below the required design minimum. The designer should specify the minimum bearing dimensions and conditions. These bearing dimensions should be shown on the erection drawings.

Careful inspection of the listed tolerances will reveal that many times one tolerance will override another. The allowable variation for one unit of the structure should not be applicable when it will permit another unit of the structure to exceed its allowable variations. Restrictive tolerances should be reviewed to ascertain that they are compatible and that the restrictions can be met. For example, a requirement which states that "no bowing, warpage or movement is permitted" is not practical or possible to achieve.

5.1.1 Bowing and Warping

Bowing and warping tolerances have an important effect on the edge match-up during erection and on the visual appearance of erected architectural precast panels, both individually and when viewed together. **Maximum permissible warpage of one corner out-of-plane of the other three corners shall be 1/16 in. per ft (5 mm per m) of distance from the nearest adjacent corner, up to 1/4 in. (6 mm) maximum after installation. Bowing shall not exceed L/360 with a maximum of 1 in. (25 mm), where L is the panel length in the direction of the bow.** The requirements for bowing and warping of panels may be overridden by tolerances for panels as installed with reference to joint widths, jog in alignment and step in face.

5.2 Product Tolerances

Product tolerances relate to the dimensions and dimensional relationships of the individual precast concrete units. They are normally established by economical and practical production considerations, and functional and appearance requirements. The actual tolerances should, however, be related to the amount of repetition, and the size and other characteristics of the precast unit. The cost of manufacturing to small tolerances decreases with increased repetition. Product tolerances are applied to physical dimensions of units such as thickness, length, width, squareness, camber and bowing (warpage), and location and size of openings.

Double tees, hollow-core slabs and solid slabs are often used for exterior facades, but are not classified as "architectural panels." Since the products listed above are manufactured the same whether they are architectural or structural products, the manufacturing accuracy for the product when used architecturally should not be expected to meet "architectural panel" tolerances. If more rigid tolerances are required, they should be clearly indicated in the contract documents.

Product tolerances should comply with industry tolerances published in Refs. 25 and 26. Tolerances for dimensions of the most common products are given in Table 5.2.

The erector shall notify the precast manufacturer immediately when out-of-tolerance units prevent proper connection or fitting so that corrective action can be taken for the remaining units.

5.3 Erection Tolerances

Erection tolerances are those required for realistic fit into the building structure. Erection tolerances are determined by the characteristics of the building structure and site conditions and are provided to achieve uniform joint and planar wall conditions. Erection tolerances should be determined on the basis of individual unit design, shape, thickness, composition of materials, and overall scale of the unit in relation to the building. The specified erection tolerances affect the work of several trades and must be consistent with the tolerances of those trades.

The recommended erection tolerances for precast concrete units are shown in Figs. 5.2(a) to (h). **Final erection tolerances shall be verified and agreed upon before erection commences and, if different from those given in this Manual, stated in writing and noted on the erection drawings.** These tolerances are guidelines only.

Because erection is equipment- and site-dependent, there may be valid reasons to vary some of the recommended tolerances to account for unique project conditions. Combining liberal product tolerances with restrictive erection tolerances may place an unrealistic burden on the erector. Thus, the designer should review proposed tolerances with precasters and erectors prior to deciding on the final project tolerances.

To minimize erection problems, the dimensions and location of the in-place structure shall be checked prior to starting precast erection, see page 6. Upon completion of member alignment, and before other trades interface with the precast concrete members, it shall be verified that the precast elements are erected within the specified tolerances.

If it is critical to the project, the tolerances which take precedence on the given project should be reflected in the contract documents and should be indicated on the project erection and production drawings.

Erection tolerances shall control the position of the individual precast members as they are located and placed in the assembled structure. The primary control surfaces or features on the precast members shall be erected to be in conformance with the established erection and interfacing tolerances. Clearances are generally allowed to vary so that the primary control surface can be set within tolerance. **Product tolerances shall not be additive to the primary surface erection tolerances.** Secondary control surfaces which are positioned from the primary control surfaces by the product tolerances are usually not directly positioned during the erection process but are controlled by the product tolerances. Thus, if the

PRODUCT TOLERANCES	PRODUCTS
Length —	
± 1/4 in. (± 6 mm)	16, 17, 18
± 1/2 in. (± 13 mm)	6, 7, 8, 9, 13, 15
± 3/4 in. (± 19 mm)	3, 5
± 1 in. (± 25 mm)	1, 2, 4, 11, 12, 14
Width —	
± 1/4 in. (± 6 mm)	1, 2, 3, 5, 6, 7, 8, 9, 12, 15, 17, 18
+ 3/8 in. (+ 10 mm)	14
+ 3/8 in.,– 1/4 in. (+ 10 mm, - 6 mm)	4
± 3/8 in. (± 10 mm)	11, 13
Depth —	
+ 1/4 in., - 1/8 in.(+ 6mm, - 3 mm)	10, 18
± 1/4 in. (± 6 mm)	1, 2, 3, 5, 6, 7, 8, 9, 12, 14, 15
+ 1/2 in., - 1/4 in. (+ 13 mm, - 6 mm)	4
± 3/8 in. (± 10 mm)	11
± 1/2 in. (± 13 mm)	13
Flange Thickness —	
+ 1/4 in., - 1/8 in. (+ 6mm, - 3 mm)	1, 2, 8, 10, 12, 15
± 1/4 in. (± 6 mm)	3, 4, 13
Web Thickness —	
± 1/8 in. (± 3 mm)	1, 8, 10, 12, 15
± 1/4 in. (± 6 mm)	2, 3
+ 3/8 in., - 1/4 in. (+ 10 mm, - 6 mm)	4
± 3/8 in. (± 10 mm)	5
Position of Tendons —	
± 1/4 in. (± 6 mm)	1, 2, 3, 4, 5, 6, 8, 9, 11, 12, 14, 15, 18
± 1/8 in. (± 3 mm)	10
Camber, Variation from Design —	
± 1/4 in. per 10 ft., ± 3/4 in. max. (± 6 mm per 3 m, ± 19 mm max.)	1, 2, 12, 15
± 1/8 per 10 ft., ± 1 in. max. (± 3 mm per 3 m, ± 25 mm max.)	4
± 3/4 in. max. (± 19 mm max.)	3
± 1/2 in. max. (± 13 mm max.)	5, 15
Camber, Differential —	
1/4 in. per 10 ft., 3/4 in. max. (6 mm per 3 m, 19 mm max.)	1, 2, 5
± 1/4 in. per 10 ft., ± 1/2 in. max. (± 6 mm per 3 m, ± 13 mm max.)	15
Bearing Plates, Position —	
± 1/2 in. (± 13 mm)	1, 2, 3, 12
± 5/8 in. (± 16 mm)	4
Bearing Plates, Tipping and Flushness —	
± 1/8 in. (± 3 mm)	1, 2, 3, 4, 12, 15
Weld Plates, Position —	
± 1 in. (± 25 mm)	
± 2 in. (± 50 mm)	7
Weld Plates, Tipping and Flushness —	
± 1/4 in. (± 6 mm)	
Haunches —	
Bearing Elevation From Bottom of Panel — ± 1/4 in. (± 6 mm)	
Relative Position of Bearing Elevation in Vertical Plane — ± 1/8 in. (± 3 mm)	
Haunch Bearing Surface Squareness Perpendicular to Applied Major Load —	
± 1/8 in. per 18 in., ± 1/4 in. max. (± 3 mm per 460 mm, ± 6 mm max.)	8, 9
± 1/8 in. (± 3 mm)	10
± 1/8 in. (± 3 mm) in short direction	6
± 1/8 in. per 12 in., ± 3/8 in. max. (± 3 mm per 300 mm, ± 10 mm max.) in long direction	6

KEY:

1 = double tee
2 = single tee
3 = building beam (rect. and ledger)
4 = I-beam
5 = box beam
6 = column
7 = hollow-core slab
8 = ribbed wall panel
9 = insulated wall panel
10 = architectural wall panel
11 = pile
12 = joist
13 = step unit
14 = sheet piling
15 = single riser bleacher slabs
16 = prison cell module – single
17 = prison cell module – double
18 = prestressed concrete panels for storage tanks

(a) For more details such as graphic descriptions of features to which tolerances apply and tolerances for sleeves, blockouts, inserts, plates, endsquareness, surface smoothness, etc., see committee reports.[25,26]

Table 5.2 Typical tolerances for precast, prestressed concrete products.

The primary control surfaces are usually as shown, although this needs to be confirmed on a job-by job-basis.

a = Plan location from building grid datum
- Structural applications .... ± 1/2 in. (± 13 mm)
- Architectural applications .... ± 3/8 in. (± 10 mm)

b = Top elevation from nominal top elevation
- Maximum low .... 1/2 in. (13 mm)
- Maximum high .... 1/4 in. (6 mm)

c = Bearing haunch elevation from nominal elevation
- Maximum low .... 1/2 in. (13 mm)
- Maximum high .... 1/4 in. (6 mm)

d = Maximum plumb variation over height of element
[element in structure of maximum height of 100 ft. (30 m)] .... 1 in. (25 mm)

e = Plumb in any 10 ft. (3 m) of element height .... 1/4 in. (6 mm)

f = Maximum jog in alignment of matching edges
- Architectural exposed edges .... 1/4 in. (6 mm)
- Visually non-critical edges .... 1/2 in. (13 mm)

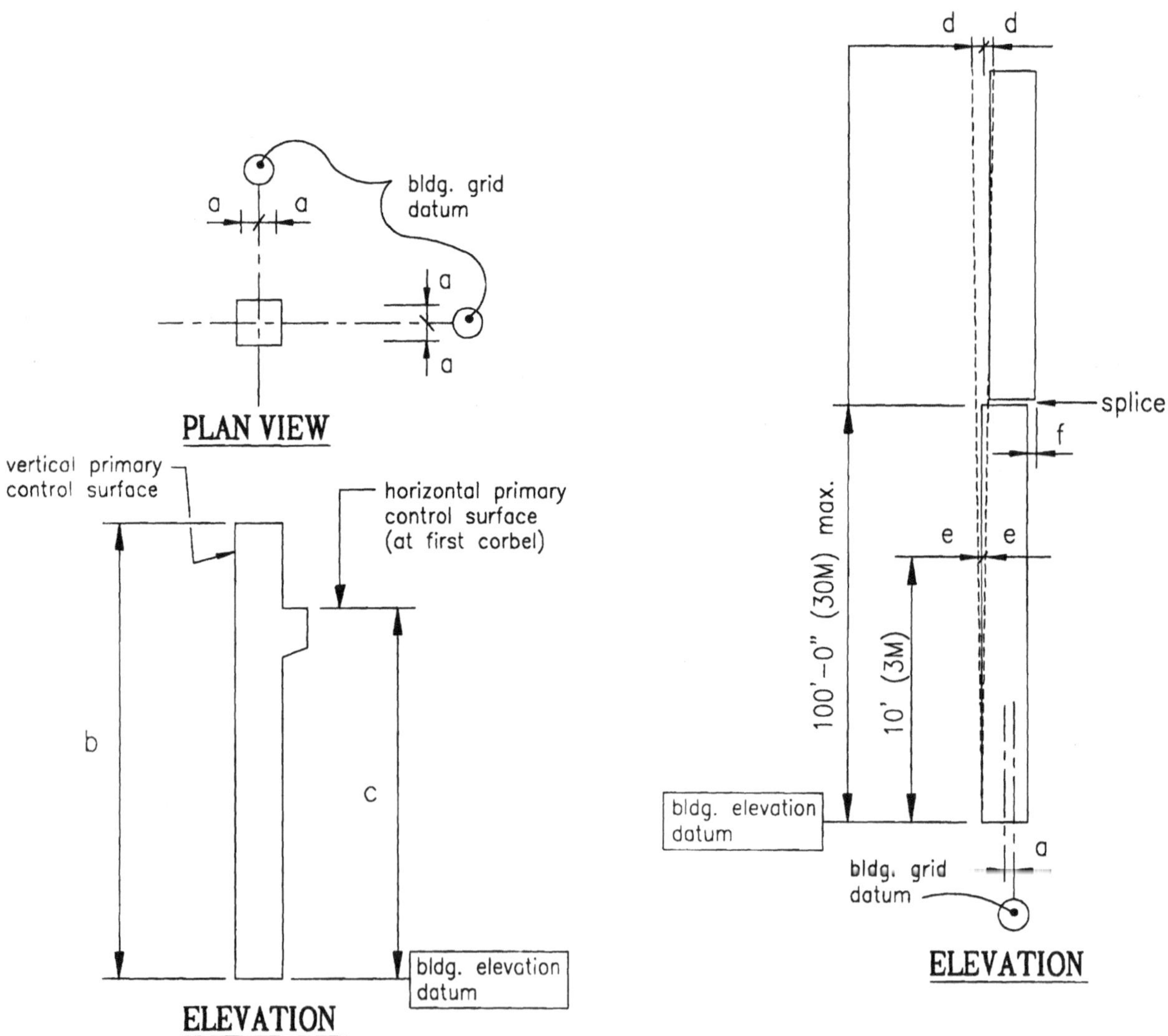

Fig. 5.2(a) Erection tolerances — columns.

The primary control surfaces are usually as shown, although this needs to be confirmed on a job-by-job basis.

a = Plan location from building grid datum .... ± 1 in. (± 25 mm)

a_1 = Plan location from centerline of steel* .... ± 1 in. (± 25 mm)

b = Bearing elevation** from nominal elevation at support
 Maximum low .... 1/2 in. (13 mm)
 Maximum high .... 1/4 in. (6 mm)

c = Maximum plumb variation over height of element
 Per 12 in. (300 mm) height .... 1/8 in. (3 mm)
 Maximum at rectangle or ell beam .... 1/2 in. (13 mm)
 Maximum at inverted tee beam .... 3/4 in. (19 mm)

d = Maximum jog in alignment of matching edges
 Architectural exposed edges .... 1/4 in. (6 mm)
 Visually non-critical edges .... 1/2 in. (13 mm)

e = Joint width
 Architectural exposed joints .... ± 1/4 in. (± 6 mm)
 Hidden joints .... ± 3/4 in. (± 19 mm)
 Exposed structural joint *not* visually critical .... ± 1/2 in. (± 13 mm)

f = Bearing length*** (span direction) .... ± 3/4 in. (± 19 mm)

g = Bearing width*** .... ± 1/2 in. (± 13 mm)

---

* For precast elements on a steel frame, this tolerance takes precedence over tolerance on dimension "a."

** Or member top elevation where a member is part of a frame without bearings.

*** This is a setting tolerance and should not be confused with structural performance requirements set by the architect/engineer. The nominal bearing dimensions and the allowable variations in the bearing length and width should be specified by the engineer and shown on the erection drawings.

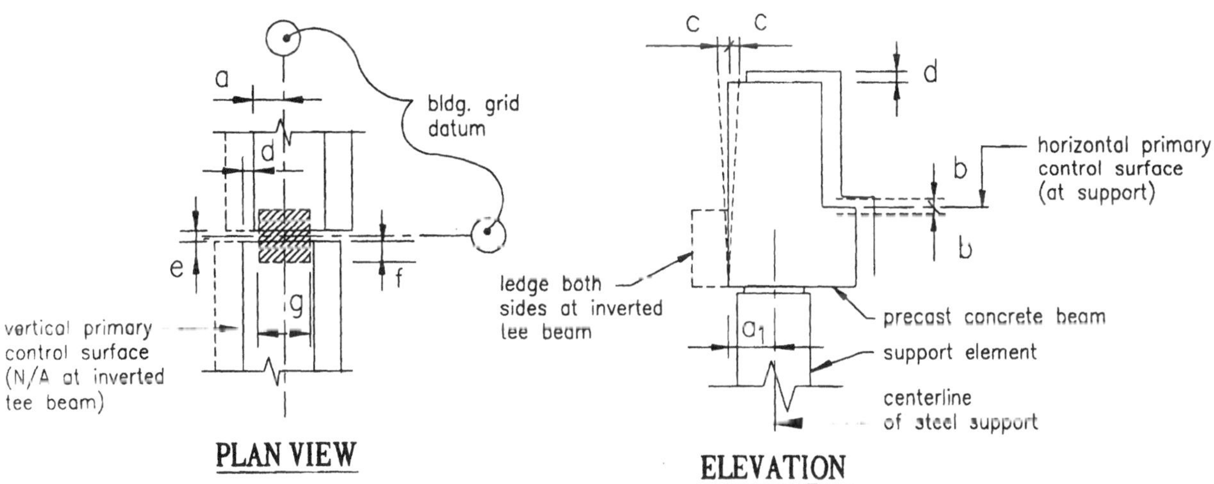

Precast element to: precast element,cast-in-place concrete, masonry, or structural steel.

Fig. 5.2(b) Erection tolerances — beams and spandrels.

The primary control surfaces are usually as shown. A majority of the time there is no vertical primary control surface, and in some scenarios there are no primary control surfaces at all. This needs to be determined on a job-by-job basis.

a = Plan location from building grid datum .... ± 1 in. (± 25 mm)

a_1 = Plan location from centerline of steel support .... ± 1 in. (± 25 mm)

b = Top elevation from building elevation datum at member ends
- Covered with topping .... ±3/4 in. (± 19 mm)
- Pretopped tee/carpet direct hollow-core .... ±1/4 in. (± 6 mm)
- Untopped roof .... ± 3/4 in. (± 19 mm)

c = Maximum jog in alignment of matching edges (both topped and untopped construction) .... 1 in. (25 mm)

d = Joint width
- 0 to 40 ft. (0 to 12 m) member length .... ± 1/2 in. (± 13 mm)
- 41 to 60 ft. (12.5 to 18 m) member .... ± 3/4 in. (± 19 mm)
- 61 ft. (18.5 m) plus .... ± 1 in. (± 25 mm)

e = Differential top elevation as erected (for units of same design and length)
- Field topped .... 3/4 in. (19 mm)
- Pretopped tees at driving lanes/carpet-direct hollow-core .... 1/4 in. (6 mm)
- Untopped roof** .... 3/4 in. (19 mm)

f = Bearing length*** (span direction) .... ± 3/4 in. (± 19 mm)

g = Bearing width*** (n/a for hollow-core) .... ± 1/2 in. (± 13 mm)

h = Differential bottom elevation of exposed hollow-core slabs**** .... 1/4 in. (6 mm)

---

* For precast concrete erected on a steel frame building, this tolerance takes precedence over tolerance on dimension "a."

** It may be necessary to feather the edges to ± 1/4 in. (± 6 mm) to properly apply some roof membranes.

*** This is a setting tolerance and should not be confused with structural performance requirements set by the architect/engineer. The nominal bearing dimensions and the allowable variations in the bearing length and width should be specified by the engineer and shown on the erection drawings.

**** Untopped installations will require a larger tolerance.

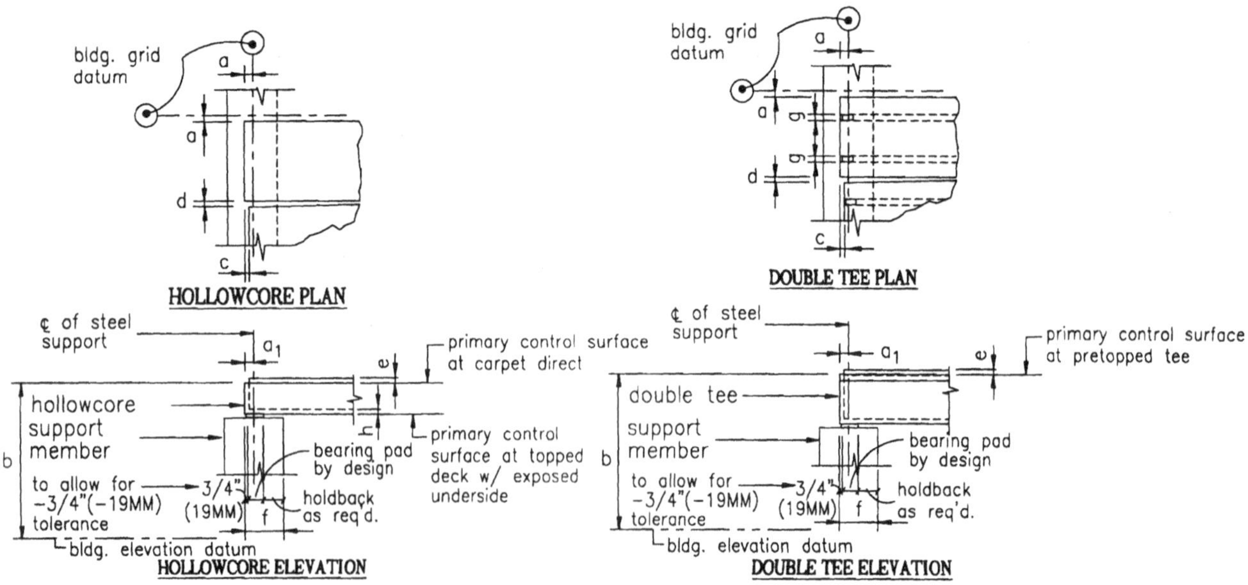

Precast element to: precast, cast-in-place concrete, masonry or structural steel support

Fig. 5.2(c) Erection tolerances — floor and roof members.

The primary control surfaces are usually as shown, although this needs to be confirmed on a job-by-job basis.

a	= Plan location from building grid datum*	± 1/2 in. (± 13 mm)
a_1	= Plan location from centerline of steel**	± 1/2 in. (± 13 mm)
b	= Top elevation from nominal top elevation	
	Exposed individual panel	± 1/2 in. (± 13 mm)
	Nonexposed individual panel	± 3/4 in. (± 19 mm)
	Exposed relative to adjacent panel	1/2 in. (13 mm)
	Nonexposed relative to adjacent panel	3/4 in. (19 mm)
c	= Bearing elevation from nominal elevation	
	Maximum low	1/2 in. (13 mm)
	Maximum high	1/4 in. (6 mm)
d	= Maximum plumb variation over height of structure or 100 ft. (30 m) whichever is less*	1 in. (25 mm)
e	= Plumb in any 10 ft. (3 m) of element height	1/4 in. (6 mm)
f	= Maximum jog in alignment of matching edges	1/2 in.(13 mm)
g	= Joint width (governs over joint taper)	± 3/8 in. (± 10 mm)
h	= Joint taper over length of panel	1/2 in. (13 mm)
h_{10}	= Joint taper over 10 ft. (3 m) length	3/8 in. (10 mm)
i	= Maximum jog in alignment of matching faces	
	Exposed	3/8 in. (10 mm)
	Nonexposed	3/4 in. (19 mm)
j	= Differential bowing, as erected, between adjacent members of the same design	1/2 in. (13 mm)

* For precast buildings in excess of 100 ft. (30 m) tall, tolerances "a" and "d" can increase at the rate of 1/8 in. (3 mm) per story to a maximum of 2 in. (50 mm).

** For precast concrete erected on a steel frame building, this tolerance takes precedence over tolerance on dimension "a."

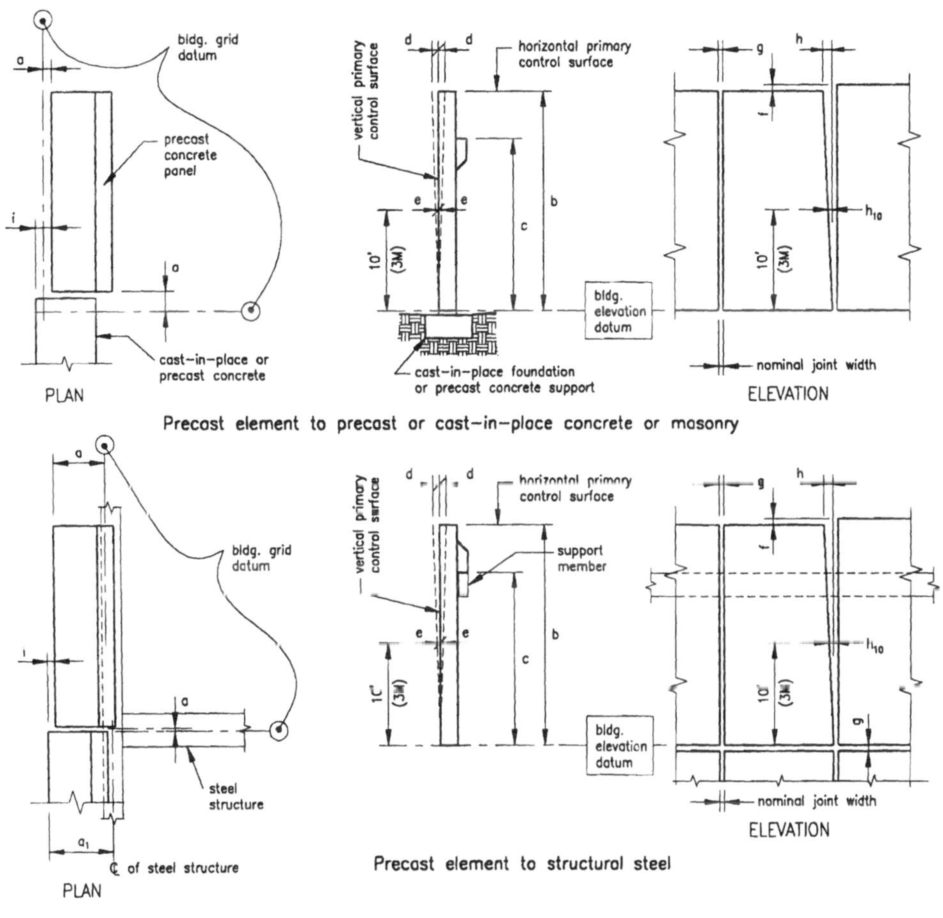

Fig. 5.2(d) Erection tolerances — structural wall panels.

The primary control surfaces are usually as shown, although this needs to be confirmed on a job-by-job basis.

a	= Plan location from building grid datum*	± 1/2 in. (± 13 mm)
a_1	= Plan location from centerline of steel†	± 1/2 in. (± 13 mm)
b	= Top elevation from nominal top elevation	
	Exposed individual panel	± 1/4 in. (± 6 mm)
	Nonexposed individual panel	± 1/2 in. (± 13 mm)
	Exposed relative to adjacent panel	1/4 in. (6 mm)
	Nonexposed relative to adjacent panel	1/2 in. (13 mm)
c	= Support elevation from nominal elevation	
	Maximum low	1/2 in. (13 mm)
	Maximum high	1/4 in. (6 mm)
d	= Maximum plumb variation over height of structure or 100 ft. (30 m) whichever is less*	1 in. (25 mm)
e	= Plumb in any 10 ft. (3 m) of element height	1/4 in. (6 mm)
f	= Maximum jog in alignment of matching edges	
g	= Joint width (governs over joint taper)	± 1/4 in. (± 6 mm)
h	= Joint taper maximum	3/8 in. (10 mm)
h_{10}	= Joint taper over 10 ft. (3 m)	1/4 in. (6 mm)
j	= Maximum jog in alignment of matching faces	1/4 in. (6 mm)
k	= Differential bowing or camber as erected between adjacent members of the same design	1/4 in. (6 mm)
l	= Opening height between spandrels	± 1/4 in. (± 6 mm)

* For precast buildings in excess of 100 ft. (30 m) tall, tolerances "a" and "d" can increase at the rate of 1/8 in. (3 mm) per story to a maximum of 2 in. (50 mm).

† For precast elements erected on a steel frame, this tolerance takes precedence over tolerance on dimension "a."

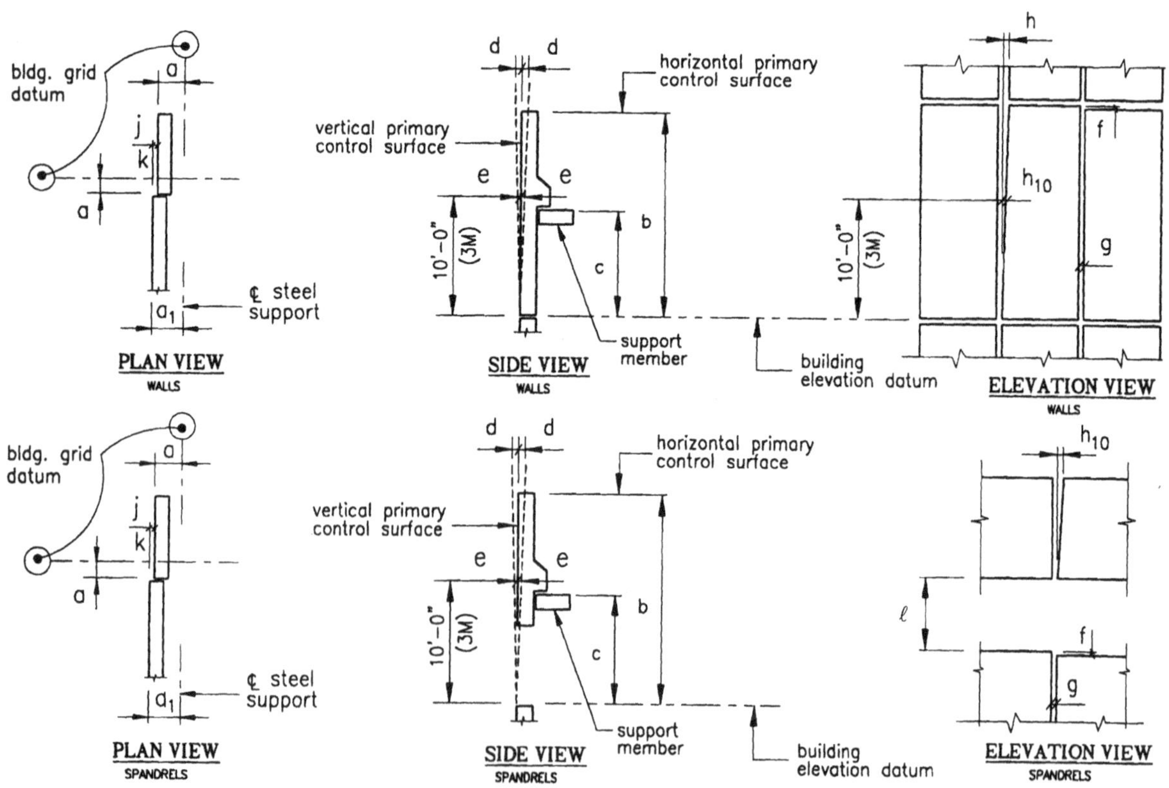

Fig. 5.2(e) Erection tolerances — architectural walls and spandrels.

The primary control surface for stair units is the top of landing at floor levels. Tolerances listed below are the same whether landings are monolithic or separate pieces.

a = Plan location from building grid line datum .... ± 1/2 in. (± 13 mm)

b = Differential elevation as erected .... ± 3/8 in. (± 10 mm)*

c = Joint width .... ± 3/4 in. (± 19 mm)

d = Maximum jog in alignment of matching edges .... 1 in. (25 mm)

e = Maximum jog in alignment of stair tread nosings .... 1/2 in. (13 mm)
(This tolerance overrides "d" if needed.)

f = Maximum jog in alignment of matching edges
at the primary control surface .... 3/8 in. (10 mm)*

g = Bearing (in span direction) .... ± 3/4 in. (± 19 mm)

---

* At stair units that have pretopped precast landings, the maximum jog between stair units as well as from stair unit to finish floor can not exceed 1/4 in. (6 mm). However, units which have landings that are topped have more leeway. This needs to be discussed and agreed upon with the general contractor.

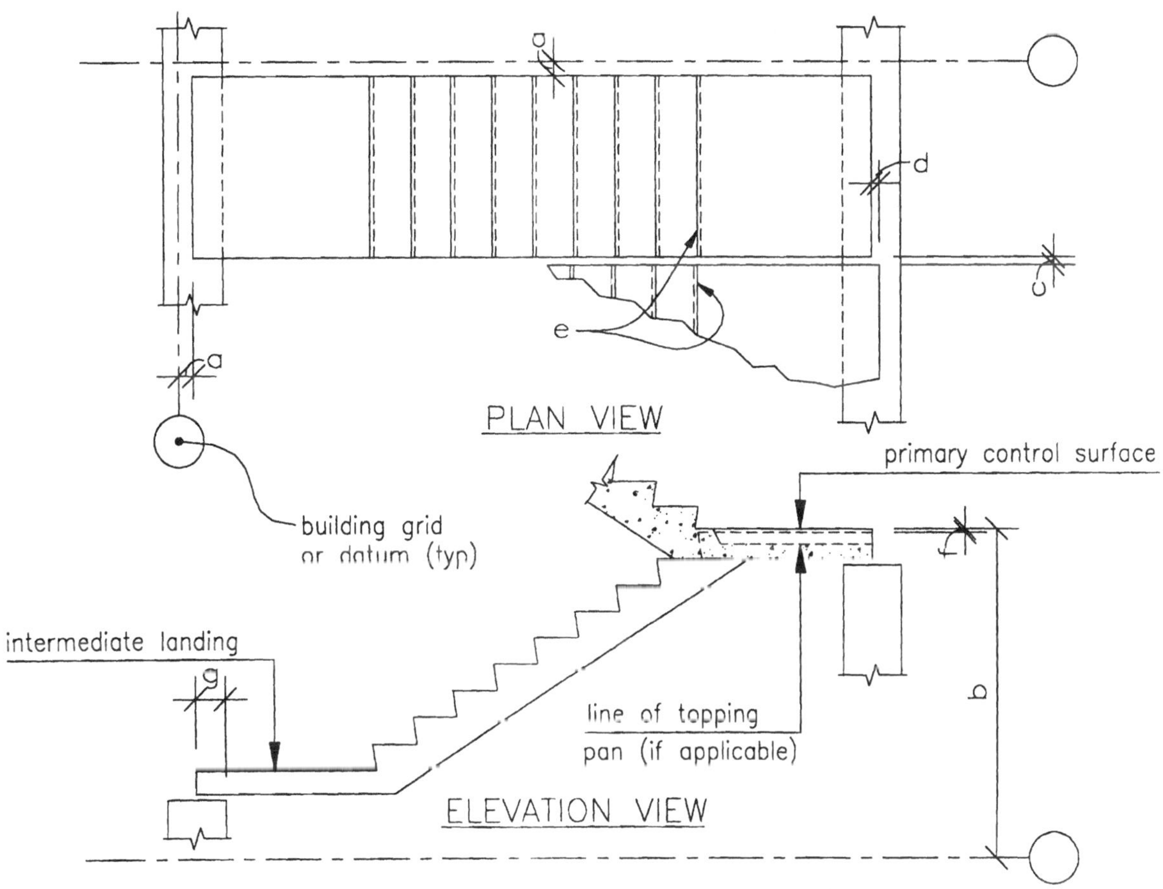

Fig. 5.2(f) Stair unit erection tolerance.

The tolerances listed below are used at the primary control surfaces only, and only those tolerances that are applicable to that surface. Normally the primary control surfaces are:

The front face of the cell unit as the vertical primary control surface, and either the head of the door (as used below), top of cell, or the bottom of balcony as the horizontal primary control surface. Note: On jobs where pretopped balconies are cast as part of the cell unit, the horizontal primary control surface may be the top of the balcony.

a = Plan location from building grid line datum . . . . ± 1/2 in. (± 13 mm)

b = Vertical control (at primary control surface) from a horizontal datum . . . . ± 3/8 in. (± 10 mm)

c = Actual grout joint . . . . 1/2 in. (13 mm) minimum

d = Plumb at element height . . . . 1/4 in. (6 mm)

e = Maximum jog in alignment of matching edges . . . . 1/4 in. (6 mm)

f = Vertical joint width . . . . ± 3/8 in. (± 10 mm)

g = Joint taper . . . . not applicable

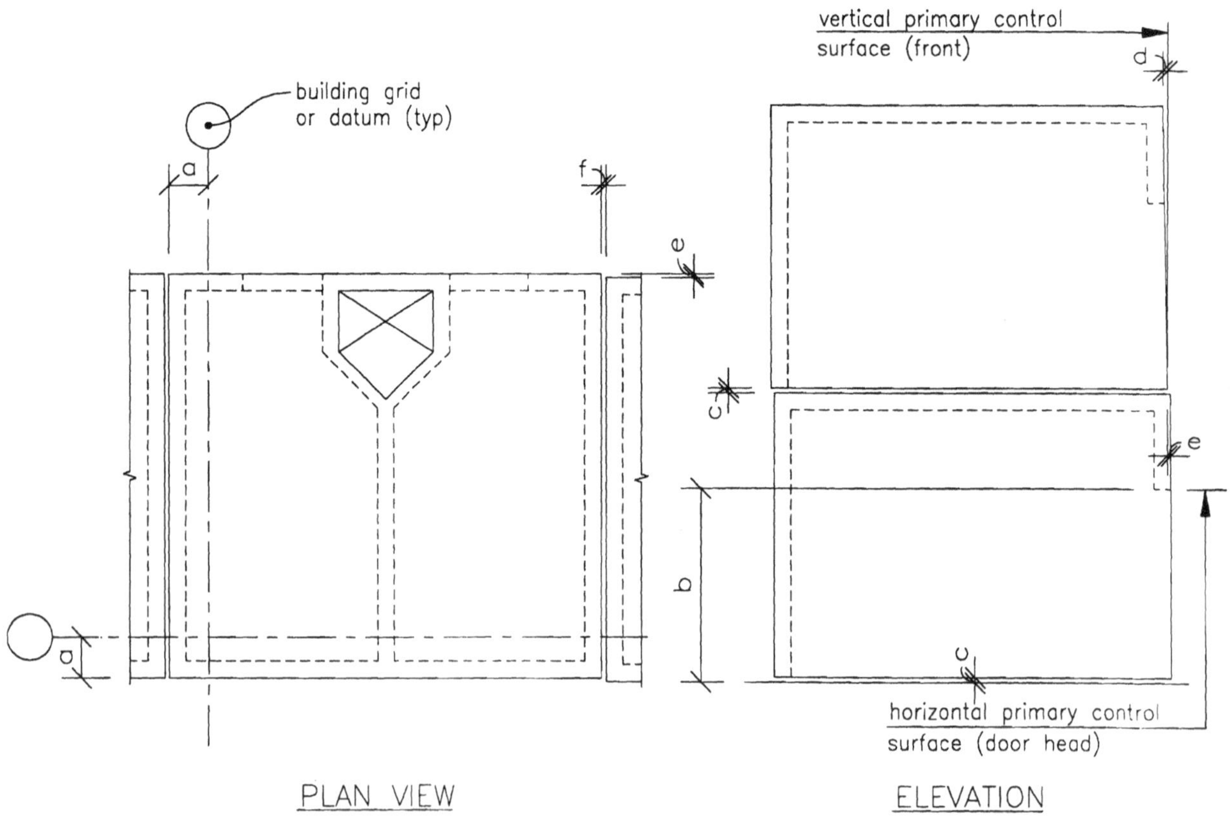

Fig. 5.2(g) Prison module erection tolerance.

The primary control surfaces are usually as shown, although this needs to be confirmed on a job-by-job basis.

a = Plan location from building grid line datum .... ± 1 in. (± 25 mm)

b = Plan location from theoretical centerline of support structure .... ± 1 in. (± 25 mm)

c = Top elevation from building elevation datum at members end .... ± 1/2 in. (± 13 mm)
This datum may be adjusted to accommodate existing field conditions.

d = Maximum jog in alignment of matching edges at the horizontal primary control surface .... 1/4 in. (6 mm)

e = Maximum jog in alignment of matching edges at the vertical primary control surface .... 1/2 in. (13 mm)

f = Bearing in span direction .... - 1 in. (- 25 mm)

g = Joint width (horizontal) at end of piece .... ± 1/2 in. (± 13 mm)
Joint needs to be 1/4 in. (6 mm) minimum.

h = Joint width (90° angle) .... ± 1/2 in. (± 13 mm)
Joint width at skewed ends .... ± 5/8 in. (± 16 mm)
Joint needs to be 1/4 in. (6 mm) minimum in either case.

j = Differential camber (at midspan as erected) between adjacent members of the same design per 10 ft. (3 m) of member length .... ± 3/16 in. (± 5 mm)

k = Differential sweep (at midspan as erected) between adjacent members of the same design per 10 ft. (3 m) of member length .... ± 3/16 in. (± 5 mm)

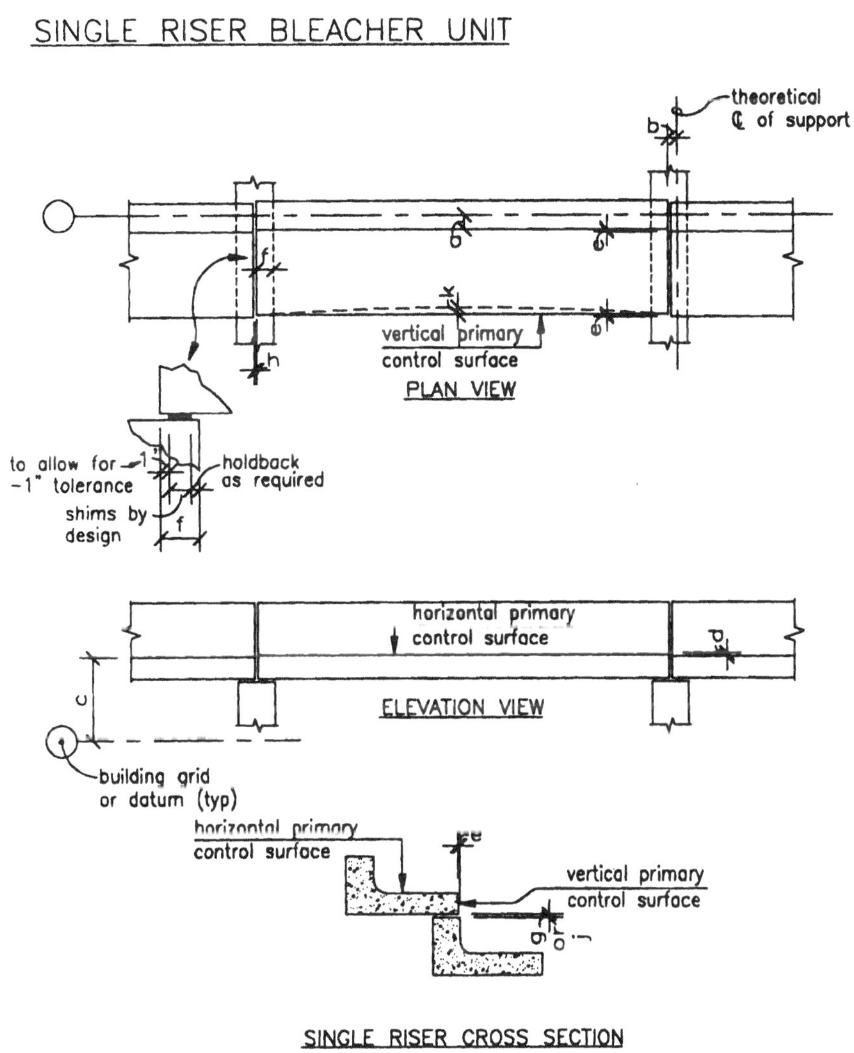

Fig. 5.2(h) Single and double riser bleacher unit erection tolerances.

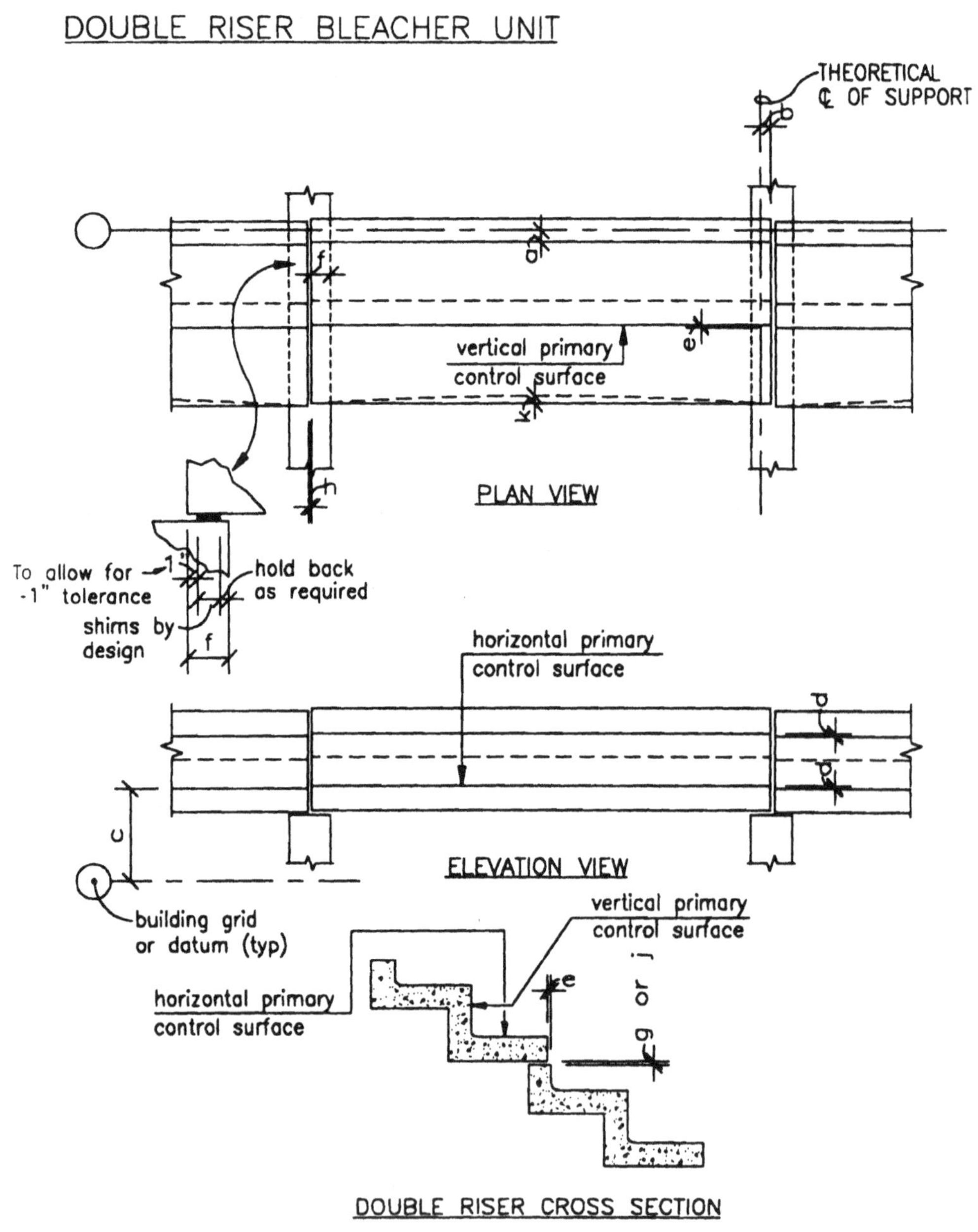

Fig. 5.2(h) (cont.) Single and double riser bleacher unit erection tolerances.

primary control surfaces are within erection and interfacing tolerances, and the secondary surfaces are within product tolerances, the member should be erected within tolerance. The result is that the tolerance limit for secondary surface may be the sum of the product and erection tolerances. To ensure trouble-free installation, the product tolerances, generally, must not exceed the erection tolerances. An example is the elevation of a second-story corbel on a multi-story column whose first-story corbel is selected as the primary elevation control surface.

Because erection and product tolerances for some features of a precast concrete member may be additive, the erection drawings should clearly define the primary erection control surfaces. If both primary and secondary control surfaces are critical, provisions for adjustment should be included. The accumulated tolerance limits may be required to be accommodated in the interface clearance. Surface and feature control requirements should be clearly outlined in the plans and specifications

If reasonable tolerances and adjustments have been designed into the construction details and followed, the erector should be able to:

1. Avoid joint irregularities such as tapered joints (panel edges not parallel), jogs at intersections, and nonuniform joint widths.
2. Maintain proper opening dimensions.
3. Properly execute all fastening connections.
4. Adjust for the accumulation of tolerances.
5. Align the vertical faces of the units to avoid out-of-plane "offsets."

During wall panel installation, priority is generally given to aligning the exterior face of the units to meet aesthetic requirements. This may result in the interior face of units being out-of-plane.

A liberal joint width should be allowed if variations in overall building dimensions are to be absorbed in the joints. This may be coupled with a closer tolerance for variations from one joint to the next for uniformity of appearance purposes. The individual joint width tolerance should relate to the number of joints over a given building dimension. For example, to accommodate reasonable variations in actual site dimensions, a 3/4 in. (19 mm) joint may be specified with tolerances of ± 1/4 in. (± 6 mm) but with only a 3/16 in. (5 mm) differential variation allowed between joint widths on any one floor, or between adjacent floors. Alternatively, the jog in alignment of edges may be specified.

The performance characteristics of the joint sealant should also be taken into account when selecting a joint size. Joints between precast concrete units should be wide enough to accommodate anticipated wall movements. When joints are too narrow, failure of the joint sealant may occur, and adjacent units may come in contact and be subjected to unanticipated loading, distortion, cracking and local crushing or spalling. If units cannot be adjusted during erection to allow for proper joint size, saw cutting may be necessary.

Variations from true length or width dimensions of the overall structure are normally accommodated in the joints or, where this is not feasible or desirable, at the corner units, in expansion joints, or in joints adjacent to other wall materials. In a situation where a joint must match an architectural feature (such as a false joint), a large variation from the theoretical joint width may not be acceptable and tolerance for building lengths may need to be accommodated at the corner units.

Erection tolerances are of necessity largely determined by the actual alignment and dimensional accuracy of the building foundation and frame. The general contractor is responsible for the plumbness, level and alignment of the cast-in-place concrete foundation and non-precast structural frame (except for precast frames) including the location of all bearing surfaces and anchorage points for the precast concrete units. If the precast concrete units are to be installed reasonably "plumb, level, square, and true," the actual location of all surfaces affecting their alignment, including the levels of floor slabs and beams, the vertical alignment of floor slab edges and the plumbness of columns or walls, must be known before the start of erection.

The architect/engineer should clearly define in the specifications the maximum tolerances permitted in the foundation and building frame alignment, then should specify that the general contractor check frequently to verify these tolerances are being held. In addition, the architect/engineer should ensure that the details in the contract documents allow for the specified tolerances. Lack of attention to these matters often necessitates changes and adjustments in the field, not only delaying the work but usually resulting in unnecessary cost, and sometimes impairing the appearance of the units and the completed structure.

Erection tolerances are less critical in structures consisting entirely of precast concrete units than for combinations of precast and cast-in-place concrete or steel frame structures. Where precast units connect to site work, such as at footings or foundation walls, large erection tolerances are particularly necessary.

Tolerances for the building frame must be adequate to prevent obstructions to member installation. The structural frame should also provide for the use of standard connections. Whenever possible, beam elevations and column locations should be uniform in relation to the precast concrete units with a constant clear distance between the precast concrete and the support elements.

In the determination of erection tolerances, attention should also be given to possible deflection or rotation of structural members supporting precast concrete. This is particularly important when bearing on flexible members such as open-web joists or cantilevered structural members. If the deflection of the structural frame is sensitive to the location or eccentricity of the connection, limits should be provided on the erection drawings. Consideration should be given to both initial deflection and to long-term deflection caused by creep of the supporting structural members.

Tolerances for a cast-in-place concrete frame are given in ACI 117[(27)] unless otherwise stated in the specifications. These tolerances are not realistic for tall buildings when one compares them to AISC tolerances and to tolerances frequently obtained on concrete buildings. Also, greater variations in heights of floors are more prevalent in cast-in-place than in other types of structures. This will affect location or mating of the inserts in the precast units with the cast-in connection devices. Tolerances for cast-in-place structures should perhaps be increased further to account for local trade practices, the complexity of the structure, post-tensioning of support structure and climatic conditions.

The following tolerances, in addition to ACI 117 requirements, should be specified when connecting precast concrete units to cast-in-place concrete structures:

1. Footings, caisson caps, and pile caps
   a. Variation of bearing surface from specified elevation: ± 1/2 in. (13mm)
2. Piers, columns, and walls
   a. Variation in plan from straight lines parallel to specified linear building lines:

      1/40 in. per ft (2 mm per m) for adjacent members less than 20 ft (6 m) apart or any wall or bay length less than 20 ft (6 m).

      1/2 in. (13 mm) for adjacent members 20 ft (6 m) or more apart or any wall or bay length of 20 ft (6 m) or more.
   b. Variation in elevation from lines parallel to specified grade lines:

      1/40 in. per ft (2 mm per m) for adjacent members less than 20 ft (6 m) apart or any wall or bay length less than 20 ft (6 m).

      1/2 in. (13 mm) for adjacent members 20 ft (6 m) or more apart or any wall or bay length of 20 ft (6 m) or more.
3. Anchor bolts
   a. Variations from specified location in plan: ± 1/4 in. (± 6 mm).
   b. Variations center-to-center of any two bolts within an anchor bolt group: ± 1/8 in. (± 3 mm).
   c. Variations from specified elevation: ± 1/2 in. (± 13mm).
   d. Anchor bolt projection -1/4 in., +1/2 in. (-6 mm, +13 mm)
   e. Plumbness of anchor bolt projection: ± 1/16 in. (± 2 mm).

Tolerances for structural steel framing should be specified to conform with the American Institute of Steel Construction (AISC) "Code of Standard Practice for Steel Buildings and Bridges," particular attention is directed to the "Commentary" included with this Code, which provides a detailed explanation of the specified erection tolerances.[(28)] Mill, fabrication and erection tolerances combined result in the final dimensional accuracy of the structural steel frame.

It should also be recognized that ACI 117 applies only to reinforced concrete buildings, and the AISC Code only to steel building frames. Neither of these standards applies to buildings of composite construction (i.e. concrete floor slabs supported by steel columns or concrete encased structural steel members). It should also be noted that fire-proofing and masonry are not controlled by reference to these standards. The precaster and erector should review the contract drawings to ensure that the location of all such materials contiguous to the precast concrete units are controlled within tolerances which are no less stringent than those specified in ACI 117. If revisions to the contract documents are indicated by this review, the architect/engineer should be notified.

The precast concrete units shall be located in the center of their theoretical location on the structure and adjusted to accommodate adjacent products, proper joint width, and alignment with adjacent precast concrete units. Small width variations between adjacent joints can be accommodated by laying-out joint centerlines spaced along an elevation and centering the units between them. The resulting joint width may vary considerably — within specified tolerance — from the theoretical width, but a smaller variation between adjacent joints should be possible.

The variations permitted for horizontal and vertical building lines should not be construed to permit extension of a building beyond its legal boundaries.

The allowable differential camber between abutted bridge deck members is usually limited to 1/2 in. (13 mm). This is an important consideration since there is often no concrete overlay to compensate for the vertical offset at the joints. However, the stiffer nature of abutted deck members leads to less total camber, and consequently less differential camber than members that receive a cast-in-place deck. Small amounts of offset between abutted deck members are normally mitigated by feathering grout across the shear key joint. Larger offsets can be minimized by shimming the girder ends to split the offset difference between the ends and midspan, or by leveling the members at midspan with a jack or lever arrangement prior to making the connection between members. **Caution shall be exercised when leveling thin-flanged abutted members, since weld plates can spall out of the thin flange under the loads imposed by leveling.**

5.4 Clearances

The clearances between the structure and precast member is an important consideration and impacts the final appearance of the structure. Clearances should be reviewed during the design stages of the project to assure they are appropriate from both erection and aesthetic points of view.

With reasonable tolerances for the building frame established, it is equally important that the designer provide adequate clearances (purposely provided space between adjacent members), for example, between the theoretical face of the structure and the back face of a precast concrete panel in detailing the panel and its relationship to the building structure. If clearances are realistically assessed, they will enable the erector to complete the final assembly without field-altering the physical dimensions of the precast units.

The failure to provide adequate clearances is an all-too-common deficiency in wall designs. They are absolutely essential for these reasons:

1. To accommodate movement between adjacent members.
2. To provide for possible size variation or misalignment.
3. To provide working space to make the connection, providing sufficient room for welding or adequate space to use a wrench to tighten bolts.

The clearance necessary for erection of the units will depend on their dimensional tolerance and geometric configuration, the dimensional accuracy of the building frame or other construction to which the units are connected and the limits of adjustment permitted by the connection details.

A rule of thumb is that at least 3/4 in. (19 mm) clearance be required between precast members, except flange-to-flange connections and wall panel to precast member where 1/2 in. (13 mm) clearance should be required with 1 in. (25 mm) preferred: 1 in. (25 mm) is the minimum clearance between the precast members and cast-in-place concrete with 2 in. (50 mm) preferred. For steel structures, 1 in. (25 mm) is the minimum clearance between the back of the precast member and the surface of the fireproofing with 1-1/2 in. (38 mm) preferred. If there is no fireproofing required on the steel, then 1 in. (25 mm) minimum clearance should be maintained. At least 1-1/2 to 2 in. (38 to 50 mm) clearance should be allowed in tall structures regardless of the structural framing material. The minimum clearance between column covers and columns should be 1-1/2 in. (38 mm), with 3 in. (75 mm) preferred because of the possibility of a column being out of plumb or a column dimension interfering with the completion of the connection. If clearances are realistically assessed, they will solve many installation problems. Where large tolerances have been allowed for the supporting structure, or where no tolerances for the structure are given, the clearance may have to be increased.

The nominal clearance dimension shown on the drawings should be equal to the actual clearance required plus the outward tolerance permitted for the adjacent construction, and should be determined based on the assumption that the construction will be as far out of position in the wrong direction as is allowed. Connections should be designed to accommodate the clearance plus the inward tolerance.

The maximum variation from the specified clearance between adjacent independent members of separated building parts or components should be ± 1/4 in. (± 6 mm). Adjacent independent members are members that are close together but are not connected structurally. These members may both be precast concrete or they may be combinations of precast concrete and cast-in-place concrete, masonry, or steel. Typically, this clearance situation develops at an expansion joint or at the interface between a stair tower and a larger structure when the two are joined only by an expansion joint.

5.5 Connection and Hardware Tolerances

All connections should be provided with the maximum adjustability in all directions that is structurally or architecturally feasible. When a 1 in. (25 mm) clearance is needed but a 2 in. (50 mm) clearance creates no structural or architectural difficulty, the 2 in. (50 mm) clearance should be selected. Closer tolerances are required for bolted connections than for most grouted connections. To accommodate any misalignment of the building frame, connections should provide for vertical, horizontal, and lateral adjustments of at least 1 in. (25 mm).

The location of hardware items cast into or fastened to the structure by the general contractor, steel fabricator, or other trades, should be located ± 1 in. (25 mm) in all directions (vertical and horizontal) from specified location, plus a slope deviation of no more than ± 1/4 in. (± 6 mm) in 12 in. (0.3 m) for critical bearing surfaces. Connection details should consider the possibility of bearing surfaces being misaligned or warped from the desired plane which would necessitate approved field adjustments to be made.

When possible, connections should be dimensioned to the nearest 1/2 in. (13 mm). The minimum clearance between various items within the connection should not be less than 1/4 in. (6 mm) with 1/2 in. (13 mm) preferred. The minimum clearance or shim space between various connection elements should be 3/4 in. (19 mm) for steel structures or 1 in. (25 mm) for cast-in-place concrete structures with 1-1/2 in. (38 mm) preferred.

When a unit cannot be erected within the tolerances assumed in the connection design, the erector shall notify the precaster and GC/CM to check the structural adequacy of the installation and the connection design should be modified, if required.

No unit shall be left in an unsafe condition. Any adjustments affecting structural performance, other than adjustments within the prescribed tolerances, shall be made only after approval by the precast design engineer. Units shall not be forced into place or installed by any method which would induce or impose any undue stress or overload on the structure, on the units, or on the connections.

5.6 Interface Tolerances

Interface tolerances and clearances are those required for the joining of different building elements in contact with or in close proximity to the precast concrete and to accommodate the relative movements expected between such elements during the life of the building. Typical examples are tolerances for window and door openings, reglets, mechanical and electrical equipment, elevators, stairs, walls and partitions.

When the matching of different building elements is dependent on work executed at the construction site, interface tolerances should be related to erection tolerances. When the execution is independent of site work, tolerances should closely match the normal manufacturing tolerances for the materials to be joined plus an appropriate allowance (clearance) for differential volume changes between materials. Fabrication and erection tolerances of other building elements must be considered in design as the precast concrete units must be coordinated with and must accommodate the other structural and functional elements comprising the total structure. Unusual requirements or allowances for interface should be stated in the contract documents.

It is important to note that interface tolerances may be very system-dependent. For example, windows of one type may have a different interface tolerance than windows of another type. If material or component substitutions are made after the initial design is complete, the responsibility for assuring that the interface tolerances are compatible logically passes to the party initiating the substitutions.

Interface erection tolerances, as an example, are required for windows between open shape precast units where they will intersect a joint between units. A similar condition often occurs where precast concrete is interspersed with glass or metal curtain wall elements, as in many precast concrete mullion projects. Close tolerances are often mandatory between the mullion and the glass or curtain wall. This condition demands additional flexibility that may be provided in the corner details. Also, any bow in spandrel panels is important if windows are to be installed between panels. Another example is variation from a plumb line of vertical openings, such as stair and elevator towers, or mechanical and pipe chases, which should be erected to the tolerances of the affected trade.

In order to ensure installation proceeds in an efficient and economical fashion, and that the various systems function as intended over the life of the structure, the erector should understand the requirements of the interfacing materials.

DIVISION 6 — QUALITY CONTROL

6.1 Philosophy

This manual is intended to be used by erectors of precast concrete as part of a quality assurance program which is either self-administered, audited by an independent agency, or both. Another reference which should be used in conjunction with this manual is the PCI "Erection Safety Manual for Precast and Prestressed Concrete" (MNL-132).[17] These manuals are the basis for audits for the PCI Field Qualification Program.

An effective quality assurance program will help the erector ensure that precast members are safely placed in accordance with the plans and specifications within the tolerances established the industry. By following the provisions in these manuals and upon developing and adhering to a program of daily inspection and reporting, the erector will have the tools to monitor the performance of the crews and thereby ensure industry standard and safe working practices.

The qualification of an erector by PCI affirms that the company and its personnel are capable of erecting precast and prestressed concrete products to industry standards. Qualification also affirms that the erector has a quality assurance program in place to monitor field operations. This process does not guarantee the quality of specific precast and prestressed concrete products in place nor is it intended to evaluate a specific project.

The evaluation of aesthetic issues such as color uniformity and finishes which are seldom under the control of the erector, while important to the final product, are beyond the scope of the PCI Field Qualification Program.

6.2 What to Inspect

The documents available on-site shall include the most recent revisions of erection drawings, sequence/load lists, all approved changes, any applicable survey and layout data, product piece weights, and when required, load lifting diagrams, and a written fall protection plan.

Other documentation when required includes welders' certifications, OSHA 200 forms, HAZCOM program [Material Safety Data Sheets (MSDS) for all materials used on site], etc., and shall be kept on file and available for review.

When a fall protection plan is required, it shall be followed. Procedures shall be observed and deviations from the plan corrected immediately. In addition, all other applicable safety requirements whether by OSHA or any other entity shall be applied.

Loose erection materials should be checked as they arrive at the site for conformance to the erection material list prior to being inventoried and stored. **Special requirements for storage of loose erection materials shall be followed.**

Materials should be kept free of dirt and other contaminants. **Grout and welding rods shall be kept dry. Verify on packages that grout is not beyond expiration dates and that packages are stored within the temperature range noted for proper mixing.**

The erector shall follow the requirements of the guying and bracing plans to maintain stability during erection. The equipment and material used shall be properly sized and in good condition. The loads to be applied to braces and guy wires shall be within their rated capacity.

The product shall be handled in accordance with the procedures outlined in Division 2 in this manual. Temporary storage, if necessary, should be on dunnage placed on firm ground capable of supporting the members without imposing unacceptable stresses which could lead to warping and cracking.

Connections shall be inspected for conformance to the erection drawings.

Bearing conditions shall be inspected. Verify the size of the shims or pads and notify the engineer if shim stacks are needed beyond the expected tolerances outlined in Division 5 in this manual. Substitution of materials shall not be permitted without the approval of the precast engineer.

Welding techniques shall follow ANSI/AWS requirements. Technique should be checked and welders certified for the types of welds required. **Compliance with welding procedure specification shall be checked.**

For grouting operations, the materials shall be verified for conformance to the specification. The

proper amount of mix water shall be monitored, and cube samples made and tested as required. Grout can be either pre-bedded, pumped or dry-packed. A procedure should be developed for each type of placement. **The grout used for proprietary connections shall be used in accordance with the manufacturer's instructions.**

Post-tensioning jacks and gages shall be calibrated. Tensioning records shall be kept. These connections shall be made and inspected in accordance with the engineer's requirements.

The precast units shall be erected within the tolerances stated in Division 5 unless superseded by the project plans and specifications. Joints should be aligned and of uniform width.

Cracks and major damage which occur during erection shall be noted and reported to the precaster. A certain amount of minor chipping and spalling should be acceptable especially if not visible in the completed structure. **Pry-bars that are used to move the members just prior to releasing the full weight on the bearings shall be placed so that any spalling is not detrimental to the durability or aesthetics of the precast.**

Jobsite corrections or modifications to connections shall be approved by the precast engineer.

Final overall structure shall be inspected for the presence of chips, cracks, warpage, stains (perhaps from other building trades) and any other detrimental defects that might affect the final appearance of the completed structure.

Erectors are usually not responsible for the final cleaning of the structure. **The erector shall make every effort to prevent jobsite soiling of the members and shall clean their work area daily of erection debris.**

6.3 Records

6.3.1 Recordkeeping

To establish evidence of proper installation and conformance with established industry standards, a system of recordkeeping shall be used to provide necessary written documentation of the precast concrete erection process. Records shall be designed for a minimum of writing.

Jobsite recordkeeping shall be the responsibility of the supervisor in charge. Records shall be kept for a minimum of 5 years after final acceptance of the structure, or for the period of product warranty provided by the manufacturer, whichever is longer. (Note: On some projects, precasters or governmental agencies may require a longer period of record retention.)

6.3.2 Records to be Kept at Jobsite

The following records and information shall be kept at the jobsite:

1. **Daily Project Erection Report. This is normally customized by the erection company and may be kept at the office. (See sample form in Fig. 6.3.2.)**
2. **Government Agency Mandated Reports including:**
   a. **Name of "Competent Person" on site.**
   b. **Job-specific written "Hazard Analysis".**
   c. **Safety Policy Manual.**
   d. **OSHA 200 Forms (these may be kept at the main office off-site).**
   e. **Crane certification document.**
   f. **Crane inspection logs.**
   g. **Erection Safety Plan (including Fall Protection Plan).**
   h. **HAZCOM Program [including Material Safety Data Sheets (MSDS)].**
3. **Connection documentation shall include the following — if applicable:**
   a. **Welder certificates.**
   b. **Welding reports, if required.**
   c. **Grouted sleeve reports, if required.**
   d. **Bolt torque, if required.**
   e. **Post-tensioning — The jacking force and actual net elongation of each tendon with allowance made for elastic shortening of the member.**

6.3.3 Product Damage and Repair Reports[29,30]

1. Defects on precast units as received.
2. Damage that occurred to units on-site.
3. Authorized Repairs.
4. **Repair records shall be kept with the "As-Built" documents.**

6.3.4 Inspector Checklist

A suggested checklist that could be used as a management tool to aid in the inspection of a structure is shown in Fig. 6.3.4. The form may be modified as needed for different projects or conditions.

Date				
Crane Size				
Number of People on Job				
General Area Erected (Give Floor and Grid Lines)				
Number of Pieces Erected				
Number of Hours Worked By Name	Hours Erecting	Hours Grouting	Hours Layout	Hours Other Work
1.				
2.				
3.				
4.				
5.				
6.				
7.				
8.				
Loads Load Number	Time Ordered	Time Arrived	Hours Job	Hours Waiting
Product Problems and Product Damage	Piece Number			
Notes:				
Near Miss Accidents and Dangerous Situations	Category			

Fig. 6.3.2. Sample daily erection report.

Fig. 6.3.4 INSPECTOR CHECKLIST — SAMPLE FORM

	N/A	Yes	No

I. PRECONSTRUCTION PLANNING

A. Field Verification and Layout

1. Erector has performed a layout of the lines and elevations to which the precast is to be erected: ____ ____ ____

   Comments: ________________________________

   ____

2. Erector has documented out of tolerance discrepancies at the jobsite and notified the precaster and general contractor: ____ ____ ____

   Comments: ________________________________

   ____

B. Determination of Erection Sequence and Direction

1. The sequence has been established and agreed to by the contractor, precaster and erector: ____ ____ ____

   Comments: ________________________________

   ____

2. The sequence has taken into consideration the stability of the structure during erection: ____ ____ ____

   Comments: ________________________________

   ____

3. Special shipping considerations have been considered ____ ____ ____

   Comments: ________________________________

   ____

C. Project Safety

1. Safety Policy and Procedures have been developed and communicated to the crew: ____ ____ ____

   Comments: ________________________________

   ____

2. Job Hazard Analysis has been made: ____ ____ ____

   Comments: ________________________________

   ____

3. Erection Safety Plan including Fall Protection Plan has been written and is available at the jobsite: ____ ____ ____

   Comments: ________________________________

   ____

C. Project Safety (cont.)

	N/A	Yes	No
4. Hazardous Material Procedures have been developed:	____	____	____

Comments: ____

____

II. PRACTICES AND PROCEDURES

A. Unloading

	N/A	Yes	No
1. Erector understands how the product will be arriving on the trailer and has made adequate provisions to unload the product:	____	____	____
a. Product can be unloaded safely and without damage:	____	____	____
b. Trailer can be positioned within the crane's reach and capacity:	____	____	____
c. Erector is documenting all deficiencies in the quality of the product upon arrival which are judged to be major and notifying the precaster:	____	____	____

Comments: ____

____

B. Access

	N/A	Yes	No
1. Erector understands the access requirements for cranes and trucks:	____	____	____
a. Access is stable and level:	____	____	____
b. There is protection of slabs, sidewalks, utilities, etc.:	____	____	____
c. Hazards and obstructions have been identified (overhead and underground):	____	____	____

Comments: ____

____

C. Site Storage

	N/A	Yes	No
1. Unloaded product has dunnage properly placed:	____	____	____

Comments: ____

____

	N/A	Yes	No
2. Dropped trailers are safely secured:	____	____	____

Comments: ____

____

D. Installation

	N/A	Yes	No
1. Erector has knowledge of the erection drawings and details:	____	____	____

Comments: ____

____

	N/A	Yes	No
2. Products are being erected in a safe manner:	____	____	____

Comments: ____

____

D. Installation (cont.)

	N/A	Yes	No
3. Bearing pads are located as shown on the erection drawings:	____	____	____

Comments: ____

____

	N/A	Yes	No
4. Members have at least the minimum bearing as shown on the erection drawings:	____	____	____

Comments: ____

____

	N/A	Yes	No
5. Erector has knowledge of the specified tolerances:	____	____	____

Comments: ____

____

	N/A	Yes	No
6. Product is being handled in accordance with precaster's recommendations to avoid damage to the product and other work associated with the product:	____	____	____

Comments: ____

____

	N/A	Yes	No
7. Products are being braced and/or connected properly and in a timely manner to ensure stability:	____	____	____

Comments: ____

____

	N/A	Yes	No
8. Products are being erected to the layout lines and elevations within tolerances:	____	____	____

Comments: ____

____

	N/A	Yes	No
9. Product and structure alignment is being monitored to ensure that there is no movement that could affect stability or tolerances:	____	____	____

Comments: ____

____

	N/A	Yes	No
10. Erector is documenting damage done to the product and notifying the precaster:	____	____	____

Comments: ____

____

	N/A	Yes	No
11. Planning and coordination of delivery schedules is ongoing. Erector is providing timely sequence lists to precaster:	____	____	____

Comments: ____

____

	N/A	Yes	No
12. Erector has documentation of discrepancies found on the erection drawings during erecting and has notified the precaster of production and detail errors:	____	____	____

Comments: ____

____

E. *Field Considerations for Connections*

	N/A	*Yes*	*No*
1. Connections are being installed per the approved erection drawings and details:	____	____	____

Comments: ____

____

2. Field modifications are approved by precast engineer:	____	____	____

Comments: ____

____

3. Connections are being installed by qualified personnel:	____	____	____
a. Welders certified in the procedure required:	____	____	____
b. Erector understands the details and how the connections are to be made:	____	____	____

Comments: ____

____

4. Erection materials for connections are on the site:	____	____	____
a. Sufficient quantities are on hand so as to not delay erection:	____	____	____
b. Erection materials, including welding rods, are stored properly:	____	____	____

Comments: ____

____

5. Expansion and chemical anchors are installed in accordance with manufacturer's recommendations:	____	____	____

Comments: ____

____

F. *Grout/Drypack*

1. Proper grout is being used per drawings and specifications:	____	____	____

Comments: ____

____

2. Grout is being stored, mixed and placed according to manufacturer's instructions:	____	____	

Comments: ____

____

G. *Post Installation Requirements*

1. Erector has completed all connections and they have been inspected and approved:	____	____	____

Comments: ____

____

2. Erector has removed all temporary bracing and connections:	____	____	____

Comments: ____

____

G. Post Installation Requirements (cont.)

	N/A	Yes	No
3. Erector has documented and notified the precaster of deficiencies in product quality or damage:	____	____	____

Comments: ______________________________

____

	N/A	Yes	No
4. Erector has picked up and removed trash, debris, leftover materials, rigging, blocking and braces:	____	____	____

Comments: ______________________________

____

	N/A	Yes	No
5. Erector has patched and repaired lifting points and connections, if required, within scope of work:	____	____	____

Comments: ______________________________

____

	N/A	Yes	No
6. Repairs and remedials are documented:	____	____	____

Comments: ______________________________

____

	N/A	Yes	No
7. Repairs are structurally and aesthetically acceptable:	____	____	____

Comments: ______________________________

____

III. EQUIPMENT

A. Cranes

1. Selection of the crane has taken into consideration the following:

	N/A	Yes	No
a. Weights of the products and safety factors:	____	____	____
b. Dimensions of the products:	____	____	____
c. Structure height:	____	____	____
d. Reach across the structure:	____	____	____
e. Lift capacity:	____	____	____
f. Mobility:	____	____	____
g. Site access restrictions:	____	____	____
h. Overhead obstructions:	____	____	____

Comments: ______________________________

____

	N/A	Yes	No
2. Condition of the crane:	____	____	____
a. The crane is well maintained:	____	____	____
b. Required inspection reports are documented and available:	____	____	____

A. Cranes (cont.)

	N/A	Yes	No
c. Capacity charts are posted or available on the crane:	____	____	____
d. The number of parts in the load line is adequate for the picks being made:	____	____	____
e. Crane is level and on stable ground:	____	____	____
f. Crane is being operated within its capacity rating:	____	____	____
g. Crane has been barricaded to stop unauthorized access at the counterweight swing:	____	____	____
h. Operator has knowledge of the erection sequence, crane set-up locations, weight of products and obstructions which may affect the safe operation of the crane:	____	____	____
i. Crane is being operated in a safe manner by a qualified operator:	____	____	____

Comments: ______________________________

____

B. Rigging

1. Selection of the rigging has taken the following items into consideration:

	N/A	Yes	No
a. Weight of the products:	____	____	____
b. Dimensions of the products:	____	____	____
c. Special conditions requiring custom rigging:	____	____	____
d. Product picking points and type of lifting anchors:	____	____	____

Comments: ______________________________

____

2. Condition of the rigging.

	N/A	Yes	No
a. Slings, spreaders, hooks, shackles, belts and rolling blocks are in good working order with no damage or defects:	____	____	____
b. Rigging has been selected that meets the manufacturer's published rating for any component of the rigging:	____	____	____
c. Special or custom handling devices has been approved by a professional engineer and erector's personnel have been instructed in proper usage:	____	____	____
d. Rigging selected fits the lifting requirement for the product being hoisted:	____	____	____

Comments: ______________________________

____

	N/A	Yes	No
e. Hoisting diagrams for non-typical units showing proper handling are available:	____	____	____

Comments: ______________________________

____

	N/A	Yes	No
f. Bolted lifting devices are checked to assure proper thread engagement:	____	____	____

Comments: ______________________________

____

C. *Operating Equipment*

	N/A	Yes	No
1. Erector has adequate equipment to perform the work:	____	____	____

Comments: ____________________

____

2. Selection of equipment has taken into consideration safety, reliability and productivity:	____	____	____

Comments: ____________________

____

D. *Personnel Access*

1. There are suitable ladders, scaffolds, and mobile elevating work platforms available:	____	____	____

Comments: ____________________

____

IV. *SAFETY OPERATIONS — Refer to Erection Safety Manual (PCI MNL-132),[17] Pages 84-95 for specific items to be inspected.*

The following safety operations have been evaluated:

A. Program Administration:	____	____	____
B. Hazard Communication (training documentation):	____	____	____
C. Housekeeping and Sanitation:	____	____	____
D. Personal Protective Equipment:	____	____	____
E. Scaffolds — General:	____	____	____
F. Swinging (Suspended) Scaffolds:	____	____	____
G. Supported Scaffolds:	____	____	____
H. Mobile Scaffolds:	____	____	____
I. Ladders:	____	____	____
J. Mobile Elevating Work Platforms:	____	____	____
K. Power Tools:	____	____	____
L. Powder-Actuated Tools:	____	____	____
M. Hand Tools:	____	____	____
N. Material Storage and Handling:	____	____	____
O. Cranes:	____	____	____
P. Rigging:	____	____	____
Q. Unloading/Lifting:	____	____	____
R. Fall Protection:	____	____	____
S. Bracing and Guying:	____	____	____

IV. SAFETY OPERATIONS (cont.)

	N/A	Yes	No
T. Connecting/Releasing Product from Crane:	____	____	____
U. Welding and Cutting:	____	____	____

Comments: ____

____

____

V. TOLERANCES (See Division 5)

	N/A	Yes	No
A. Erection	____	____	____
B. Interfacing	____	____	____
C. Clearances	____	____	____

REFERENCES

1. "Rigging Qualifications and Responsibilities in the Construction Industry," ANSI A10.42, American National Standards Institute, New York, NY (to be published).
2. *Rigging Manual,* Construction Safety Association of Ontario, Toronto, Ontario, Canada, 1996, 118 pp.
3. Rossnagel, W. A., MacDonald, J. A., and Higgens, L. R., *Handbook of Rigging for Construction and Industrial Operations,* McGraw-Hill, Inc., New York, NY, 1988.
4. *PCI Design Handbook, Precast and Prestressed Concrete,* Fifth Edition, Precast/Prestressed Concrete Institute, Chicago, IL, 1998.
5. "Tilt-Up Concrete Association's Guideline for Temporary Wind Bracing of Tilt-Up Concrete Panels During Construction," TCA Guideline 5-98, Tilt-Up Concrete Association, Mount Vernon, IA, 1998, 11 pp.
6. *Design and Typical Details of Connections for Precast and Prestressed Concrete (MNL12388),* Second Edition, Precast/Prestressed Concrete Institute, Chicago, IL, 1988, 270 pp.
7. PCI Committee on Connection Details, "Addendum to Design and Typical Details of Connections for Precast and Prestressed Concrete," PCI JOURNAL, V. 40, No. 5, September-October 1995, pp. 46-57.
8. *Structural Welding Code-Steel,* D1.1-98, American Welding Society, Miami, FL, 1998, 440 pp.
9. Joint PCI-PTI Committee on Segmental Construction "Recommended Practice for Precast Post-Tensioned Segmental Construction," PCI JOURNAL, V. 27, No. 1, January-February 1982, pp. 14-61.
10. "Specification for Unbonded Single Strand Tendons," Revised 1993, Post-Tensioning Institute, Phoenix, AZ, 1993, 20 pp.
11. "Recommenced Practice for Grouting of Post-Tensioned Prestressed Concrete," *Post-Tensioning Manual,* Fifth Edition, Post-Tensioning Institute, Phoenix, AZ, 1990, pp. 230-236.
12. ACI Committee 301, "Standard Specifications for Structural Concrete (ACI 301-96)," American Concrete Institute, Farmington Hills, MI, 1996, 43 pp.
13. "Test Method for Time of Setting of Hydraulic Cement by Vicat Needle (ASTM C191)," ASTM, West Conshohocken, PA.
14. *PCI Bridge Design Manual (MNL-133-97) – Chapter 3: Fabrication and Construction,* Precast/Prestressed Concrete Institute, Chicago, IL, 1997.
15. *Crane Handbook,* Construction Safety Association of Ontario, Toronto, Ontario, Canada, 1975, 304 pp.
16. *Mobile Crane Manual,* Construction Safety Association of Ontario, Toronto, Ontario, Canada, 1982, 419 pp.
17. *Erection Safety Manual for Precast and Prestressed Concrete (MNL-132-95),* Precast/Prestressed Concrete Institute, Chicago, IL, 1995, 120 pp.
18. Fisher, Timothy S., "Choosing Mobile Cranes for Erecting Precast Members," *Concrete Journal,* Addison, IL, March 1996, pp. 161-168.
19. "Cranes-on-Rail Lifting Scheme Makes Mass Transit Easier," *Engineering News Record,* New York, NY, October 6, 1997, pp. 26-28.
20. Yee, Alfred A., and Masuda, Fred R., "Design for Erection Considerations," PCI JOURNAL, V. 19, No. 6, November-December 1974, pp. 16-27 and V. 20, No. 1, January-February 1975, pp. 74-97.
21. "Lifting System Turns Corners," *Engineering News Record,* New York, NY, July 22, 1982, pp. 17-18.
22. "Boom Supported Elevating Work Platforms (ANSI/SIA A92.5), American National Standards Institute, New York, NY, 1992.
23. *Construction Health & Safety Manual,* Construction Safety Association of Ontario, Toronto, Ontario, Canada, 1998, 344 pp.
24. *"National Electrical Code (NFPA-70),"* National Fire Protection Association, Quincy, MA, 1996.
25. PCI Committee on Tolerances, "Tolerances for Precast and Prestressed Concrete," PCI JOURNAL, V. 30, No. 1, January-February 1985, pp. 26-112.
26. PCI Committee on Tolerances, "Product Tolerances for Precast and Prestressed Concrete," PCI JOURNAL, V. 38, No. 1, January-February 1993, pp. 16-27.
27. "Standard Specifications for Tolerances for Concrete Construction and Materials and Commentary (ACI 117-90)," American Concrete Institute, Farmington Hills, MI, 1990.
28. *Code of Standard Practice for Steel Buildings and Bridges,* American Institute of Steel Construction, Chicago, IL, June 10, 1992.
29. PCI Committee on Quality Performance Criteria, "Fabrication and Shipment Cracks in Prestressed Hollow-Core Slabs and Double Tees," PCI JOURNAL, V. 28, No. 1, January-February 1983, pp. 18-39.
30. PCI Committee on Quality Performance Criteria, "Fabrication and Shipment Cracks in Precast or Prestressed Beams and Columns," PCI JOURNAL, V. 30, No. 3, May-June 1985, pp. 24-49.
31. *PCI Drafting Handbook – Precast and Prestressed Concrete,* Second Edition (MNL-119-90), Precast/Prestressed Concrete Institute, Chicago, IL, 1990, 332 pp.
32. Miller, Duane K., "What Every Engineer Should Know About Welding," *Modern Steel Construction,* Chicago, IL, May 1997, pp. 38-48.

33. Holdren, R. L., "What Are the Causes of and Solutions to Weld Quality Conflicts?," *Welding Journal,* American Welding Society, Miami, FL, August 1993, pp. 57-65.

34. *Structural Welding Code – Reinforcing Steel,* D1.4-98, American Welding Society, Miami, FL, 1998, 45 pp.

35. "Practice for Repair of Damaged and Uncoated Areas of Hot-Dip Galvanized Coatings (ASTM A780)," ASTM, West Conshohocken, PA.

36. Livelli, Gregory and Langill, Thomas, "Guidelines for Welding Galvanized Steel," PCI JOURNAL, V. 43, No. 3, May-June 1998, pp. 40-48.

37. "Welding of Stainless Steels and Other Joining Methods," Nickel Development Institute, Toronto, Ontario, Canada, 1988, 47 pp.

38. "Test Method for Sulfur in Petroleum Products (General Bomb Method) (ASTM D129)," ASTM, West Conshohocken, PA.

39. American Welding Society, Miami, FL
"A5.1 Specification for Carbon Steel Electrodes for Shielded Metal Arc Welding,"
"A5.5 Specification for Low Alloy Steel Electrodes for Shielded Metal Arc Welding,"
"A5.6 Specification for Covered Copper and Copper Alloy Arc Welding Electrodes,"
"A5.7 Specification for Copper and Copper Alloy bare Welding Rods and Electrodes,"
"A5.20 Specification for Carbon Steel Electrodes for Flux Cored Arc Welding,"
"A5.28 Specification for Low Alloy Steel Electrodes for Gas Shielded Metal Arc Welding,"
"A5.29 Specification for Low Alloy Steel Electrodes for Flux Cored Arc Welding,"

40. "Specification for Latex Agents for Bonding Fresh to Hardened Concrete (ASTM C1059), ASTM, West Conshohocken, PA.

41. "Specification for Epoxy-Resin-Base Bonding Systems for Concrete (ASTM C881), ASTM, West Conshohocken, PA.

42. ACI Committee 224, "Causes, Evaluation and Repair of Cracks in Concrete Structures (ACI 224.1R)," American Concrete Institute, Farmington Hills, MI, 1993, 22 pp.

43. ACI Committee 503, "Use of Epoxy Compounds with Concrete (ACI 503R)," American Concrete Institute, Farmington Hills, MI, 1993, 28 pp.

44. "Standard Guide for Use of Joint Sealants (ASTM C1193)," ASTM, West Conshohocken, PA.

45. "Removing Stains and Cleaning Concrete Surfaces (IS214 TC)," Portland Cement Association, Skokie, IL, 1988, 16 pp.

46. Pfeifer, W. D., and Scali, M. J., "Concrete Sealers for Protection of Bridge Structures," National Cooperative Highway Research Program, Report 244, Transportation Research Board, National Research Council, Washington, DC, December 1981.

GENERAL

Shapiro, Howard I., Shapiro, Jay P., and Shapiro, Lawrence K., *Cranes and Derricks,* McGraw-Hill, Inc., New York, NY, 1991, 480 pp.

"Crane Safety on Construction Sites," ASCE Manual and Reports on Engineering Practice No. 93, American Society of Civil Engineers, Reston, VA, 1998, 137 pp.

APPENDIX A — GLOSSARY*

Banding (bands) — steel strips pulled taut over a load and secured with crimped metal clips. Used in lieu of chains and binder.

Barring — to move or adjust a member with a steel pry bar.

Belt — Commercially manufactured cloth, synthetic nylon web, or wire mesh strip with eyes at each end, utilized when a wire rope sling could damage the member.

Binder (load binder) — a lever-actuated device used to tighten chain. Most often used on transport vehicles to secure loads.

Bloodborne pathogens — pathogic microorganisms that are present in human blood and can cause disease in humans. These pathogens include, but are not limited to, hepatitis B virus (HBV) and human immunodeficiency virus (HIV).

Bolster — A rocker-like device used to support precast concrete members during transportation.

Boom — the structural member of a crane or derrick used to carry the weight of the load and to project the upper end of the hoisting tackle in reach or in a combination of height and reach.

Bull wheel — a large diameter, horizontally mounted wheel situated at the bottom of the derrick mast and boom. Wire ropes are wound around the wheel and attached to winches at each end. Used to swing the derrick boom.

Bunk — a steel or wooden cross-member placed on a trailer or tractor fifth wheel to distribute the load of a member during transport.

Caulking — see sealant

Chicago boom — a boom which is attached to a structure, an outside upright member of the structure serving as the mast, and the boom being stepped in a fixed socket clamped to the upright.

Chain fall — Hand operated lifting device employing a continuous pull chain to raise and lower the hook.

Choker — sling or wire rope with eyes spliced on each end used to lift load; sling is placed around the load and attached back to itself forming a noose that tightens on the load.

Closed link (master link) — an alloy steel, oblong ring, commonly used between more than one sling and the load block.

Come-a-long — a lever and ratchet-operated pulling device that uses a wire rope or chain to exert a tensile load on an object.

Contractor's hardware (miscellaneous iron) — items to be placed on or in the structure in order to receive the precast concrete units; e.g., anchor bolts, angles, or plates with suitable anchors.

Dead end — Fixed end of a live hoisting line.

D-ring — the swiveling portion of a lifting plate to which the hoisting sling is attached.

Dollie — axled apparatus used to support the back end of a long precast member during the hauling process.

Dunnage — materials, usually wood, used for keeping concrete units from touching each other, the ground, or other materials during storage and transportation.

Engineer of record — engineer who develops original building design and is responsible for the design of the building or structure of which the precast concrete forms a part.

Erection hardware — all loose hardware necessary for the installation and final connections of the precast concrete units.

Equalizer beam — a structural rigging component used to equalize or distribute sling loads in a desired proportion or between cranes in a dual lift. (see Fig. below).

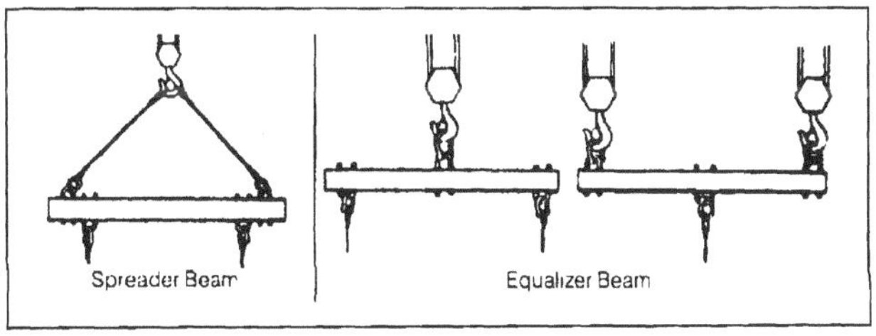

---

* For additional "Glossary" terms definitions, meanings, and uses, refer to "A Glossary of Common Crane & Rigging Terms," Specialized Carriers & Rigging Foundation, Fairfax, VA, 1997, 67 pp.

Feather — to taper mismatched elevations of two abutting precast members with grout.

Flemish eye — a method of forming an eye splice in the end of a wire rope that provides 100% efficiency when used with a pressed metal (swaged) sleeve.

Genie hoist — a small electric, gear reduced, vertical winch, most often used on swing stages.

Guy derrick — a derrick whose boom is supported by wire ropes running between the boom tip and the top of a mast supported by guy wires running from the mast tip and anchored into the supporting surface.

Jib — an auxiliary extension to the boom mounted at or near the boom tip. Can be mounted in line with the boom axis or offset at an angle to the boom axis.

Lattice boom — a boom constructed of three or four corner members (chords) assembled with transverse and diagonal members (lacings or lattices) to form a trusswork. The chords carry the axial boom forces and the bending moments while the lacings resist shear.

Lay — signifies the direction of rotation of the strands in a wire rope.

Lay length — distance measured along a rope in which a strand makes one complete revolution about the rope axis.

Lifting anchors — cast-in devices used to lift precast units.

Lifting hardware — metallic inserts in or attachments to lifting anchors to facilitate handling.

Load block — an assembly of a hook or shackle plus swivel, bearings, sheaves, pins and frame suspended at the lower end of a reeved system of hoisting lines.

Luffing — a term used with derricks to indicate raising or lowering the boom which changes the boom angle with respect to horizontal.

Material safety data sheets (MSDS) — written or printed material containing information concerning a hazardous chemical.

Miscellaneous iron — see Contractor's hardware.

Mobile elevating work platforms (MEWP) — hydraulically operated personnel carrier, often called "aerial lifts" or "manlifts."

Part of line — indicates the number of wire ropes supporting a load to provide mechanical advantage to a system.

Pick — Individual lift made with a crane.

Pick point — location of individual lifting anchors.

Primary control surface — a surface on a precast element, the dimensional location of which is specifically set and controlled in the erection process.

Rebar — reinforcing steel bar.

Reeve — to pass a line through an opening or around a pulley or sheave; the process of passing cable through sheaves to create multipart lines.

Rendering — the movement of a cable through a rolling block to equalize the force on each lift point.

Rolling block (snatch block) — a single or double sheave block arranged so that one or both cheek plates can be opened permitting the block to be reeved without having to pass a free rope end through it.

Sealant (joint sealant) — material used to exclude water and solid foreign materials from joints between precast members and such units and adjacent materials.

Sealer — a clear chemical compound applied to the concrete surface for the purpose of improving resistance to water penetration or for improving weathering qualities.

Secondary control surface — a surface on a precast element, the dimensional location of which is dependent on the location tolerance of the member primary control surfaces, plus the member feature tolerances.

Shackle — a U-shaped piece of metal with holes in each end through which a pin is placed [see Fig. 2.2.6(a)].

Sheave — a grooved wheel used to accept and support a wire rope. Used to change the direction of a wire rope.

Sling — a system of wire ropes, chains, fiber ropes or web straps used to securely attach a load to a crane hook.

Slug — bar used to fill gap between weld plates.

Snatch block — see Rolling block.

Softener — non-metallic material placed between precast concrete and other material and used as a cushion to prevent damage to precast concrete or rigging.

Spreader beam/bar/strut — a structural rigging fitting used to spread the legs on a sling to provide vertical loading only on a load being lifted.

Spreader set — a pair of slings suspended from a load block to provide multiple point attachment to a load.

Stiffleg derrick — a derrick whose boom is supported by wire ropes running between the boom tip and the top of a structural mast supported by two structural legs running from the mast tip to the supporting surface.

Story pole — a gauge normally used to measure window opening.

Strongback — a temporary structural beam or truss bolted or welded to the back of a precast member to stiffen or reinforce it during shipping and handling operations.

Swivel plate — a specialized, alloy steel device which bolts to anchors cast-in the member and includes a D-ring for the attachment of the hoisting sling.

Tagline — a rope attached to a precast concrete unit during erection for use in helping the crane operator guide the members into place.

Tandem lift — using two units of hoisting equipment to lift a single precast member.

Two-blocking — the jambing of a travelling load block or ball into a stationary hoisting sheave.

Trip — rotating a precast member using two hoist lines.

Whip line — auxilliary hoist line.

APPENDIX B — FACTORS AFFECTING ESTIMATING PROCEDURES AND ERECTION COSTS

Estimating the cost of erection of precast concrete requires knowledge and experience that can only be learned by on-the-job exposure to the actual work, as well as familiarity with the cost of labor, equipment and materials. An experienced estimator will assess all factors which influence the cost of erection, such as access conditions on the site, complexity of connections and weather problems.

The following information is desirable for precast concrete erection estimating:

1. **Name** of project, architect, engineer, general contractor or construction manager, and precast concrete manufacturer(s).
2. **Quantities,** sizes and weights of precast concrete units.
3. **Quantities** and types of connection hardware and who is responsible for supply.
4. **Schedule** of when precast units are available and in what quantity so that the erection work schedule may be determined taking into consideration the starting point and the number of move-ins.
5. **Site Inspection** (using the site plan) to check for:

   Site Access — for truck and crane. Space available for crane assembly as well as disassembly. Responsibility for site protection and relocation of overhead lines.

   Traffic — necessity for flagmen and barricades. Truck space available at site for unloading.

   Crane Set-Up Locations — determine any restrictions to crane operations.

   Ground Conditions — determine any restrictions to crane and load movements around site, such as underground services, tunnels or ramps.

   Interior Access — the need to cross footings, grade beams, buried ducts or tanks. The size of wall opening necessary and if there is excavation or construction scheduled which may interfere with access.

   Governing Restrictions and Permits — determine local restrictions regarding crane and truck movements, and licensing requirements.
6. **Sequence** — special sequencing may be required or it may be open to erector's option.
7. **Types of Connections** — whether welded, bolted or grouted.

After assembling as much of the above information as possible, the estimator should look over the plans and take-off sheets to get an overall scope of the project. In preparing the full estimate, the following should be considered:

Trucking Considerations:

- can the precast concrete units be delivered in the quantity and sequence required?
- is specialized equipment required and available?
- is a staging area for loads desirable and if so, who pays rent, if any, and provides shuttle service?
- is on-site access to unloading areas provided?
- are there travel restrictions due to height, width, length, or weight, daylight hours, weather or road conditions?
- is traffic control necessary?
- is rail delivery available?
- do special handling requirements dictate orientation and position of precast concrete on trailer (consider the problem of two-line turning required for tower crane erecting)?
- curfews in effect?

Erection Considerations:

- what are manpower requirements, availability and cost (local labor practices)?
- are layout, layout marks required (how accessible, accurate)?

- what sequence of erection must be followed to maintain building stability; which unit to start with and which way to proceed?
- who is responsible for erection structural analysis?
- how many trips are required with crane and crew to complete the project (number of move-ins)?
- how many total units are to be erected and how many units of each type can the crew reasonably expect to set, *on the average,* per day?
- what size crane will handle the most difficult pick (combination of weight and reach)?
- can precast units be erected directly from truck or is double handling and temporary on-site storage necessary?
- is hoisting equipment supplied by others available (off-shift working hours or regular hours)?
- what equipment (tools, materials, and jigs), beside the crane and aerial work platforms, will be required to efficiently and safely erect the project?
- what are working restrictions due to job location (no working from street, time limitations, power lines)?
- what time of year will job operate and what type of weather is anticipated (manpower efficiency, access conditions)?
- can connections be adjusted after release of the crane, are temporary connections needed?
- how many levels can be erected on temporary connections before final connections are made?
- welding requirements (pre-welding, pre-heating, shelter, protection of insulation), position and type of welding?
- who is responsible for adjusting or feathering surfaces between adjacent floor or roof slabs?
- is there any grouting required and, if so, who furnishes and installs?
- who provides heating and winter protection for grouting and mortaring operations, if required?
- who provides lights, power, water and how much of each is required?
- what is patching, finishing, caulking and clean-up allowance?

Contract Considerations:

- are prevailing wage rates required?
- what are schedule requirements and is there a liquidated damage clause?
- is a performance bond required and if so, who absorbs its cost?
- are there applicable taxes and who absorbs their cost?
- is there anything unusual regarding progress payments and retention?
- is change-order procedure established?
- what constitutes acceptance or rejection of the project, and is there an acceptance procedure by owner or his representatives for the completed work?
- is there Builders Risk Insurance and what is the deductible?
- who does layout work and to what extent?
- are special safety considerations required, such as 100% tie-off, fire watch, and safety director?
- what are erector responsibilities regarding OSHA compliance on this project (who erects perimeter protection and covers floor and roof holes)?
- is protection of the work after erection required and if so, who supplies and who installs?
- are there any special minority labor requirements to fulfill?
- is participation in composite crew cleanup required?

All the above questions should be answered by reviewing the contract documents, including the architect's plans and specifications, and by relying on experience. Having answered them, an estimate can be compiled and a formal proposal with cost quotation prepared.

After completing the erection estimate, clearly delineate all the special conditions and assumptions on which the estimate is based and include these conditions in the job proposal. It is extremely important to have these conditions become part of the contract as they could greatly affect erection costs.

APPENDIX C — ERECTION DRAWINGS

The primary function of precast concrete erection drawings is the translation of contract documents into usable information for accurate and efficient handling and erection of precast concrete units. Erection drawings, when submitted for approval, allow the architect or engineer to check whether the contract documents have been properly interpreted.

Erectors use the erection drawings to determine crane requirements, setting sequence with the GC/CM and precast concrete manufacturer, and to schedule work.

These drawings should show all elevations, plans and sections that are necessary to clarify conditions sufficiently to efficiently erect the structure. They must be easy to read, clear with respect to connections and explicit in instructions of sequence, if necessary.

Early coordination of design, manufacturing and erection is essential for overall economy of precast concrete construction. Before drawings are started, the precast concrete manufacturer's engineering personnel should meet with the production, shipping and erection personnel to discuss overall job procedures such as panel sizes and connection details. At this time it may be advisable to meet with the GC/CM, architect and engineer to discuss and agree upon any desired changes. It is important that a record of information and decisions made in the pre-drawing stages be distributed to all concerned parties.

Normal practices for the preparation of drawings for precast concrete are described in detail in the *PCI Drafting Handbook — Precast and Prestressed Concrete.*[31] However, the following is a brief synopsis of the information required to give the erection personnel adequate information in order to erect the precast concrete correctly and efficiently.

Cover Sheet

1. *General*
   The name of the precaster's chief or project engineer and the telephone number of the engineering department. The project name, as it appears on the architectural plans, and the precaster's job number.

2. *Site Map*
   A sketch of the jobsite indicating major traffic arteries and significant landmarks to aid in the location of the site.

3. *General Notes*
   Specify welding requirements and minimum bolt penetrations for erection. Note whether units are furnished erected or F.O.B. Specify contract inclusions and exclusions. Specify engineering requirements such as cast-in-place concrete strength, special stability requirements of structure and individual members, minimum bearing, etc. List the precast concrete manufacturer's requirements of other sub-contractors. Indicate other special conditions, such as inspection requirements, particularly for welding.

4. *Erection Drawing List*
   A complete list of all erection drawings should be included on the cover sheet. Sheet number and description should be indicated.

5. *Blocking Requirements*
   Unless otherwise noted, blocking is at lifting points.

Key Plan

A key plan is recommended on each sheet for any project requiring more than one sheet of elevations or erection plans. This plan gives the reader a common reference and is especially useful on a project where an extremely complicated building, or a number of separate buildings, is involved. A key plan is also useful whenever a general job discussion arises. Such a plan should include:

1. Major grid and column lines and dimensions, both intermediate and overall.

2. Indication on which sheet each part of the building appears in elevation or plan.

3. Construction north and true north, if different.

Schedules

Schedules are a summary of all precast products on a particular project and become a valuable quick reference tool for the erector. These schedules should contain the following information:

1. A general description of the type of precast unit (i.e., 18 in. double tee, 6 in. thick architectural panel, etc.). This can also include a basic cross-sectional picture.
2. The overall size and weight of each unit, along with mark number.
3. The quantity of each mark number type.

This information will help the erector to schedule the loads in the most efficient manner, as well as plan the handling and sequencing for each unit during erection. The schedule is also a place to check-off units as they are delivered and erected.

Elevations

Elevation drawings are the precaster's graphic definition of architectural and structural intent based on the contract documents. They must be clear enough to be checked easily by the architect, engineer and GC/CM, and must contain all information vital to the shipping or erection process. Elevation drawings should include:

1. Precast unit mark number. Any members which require sequential erection should be properly indicated.
2. Width and height of each member.
3. Location of precast concrete units to floor elevations, top of structural support and grid lines.
4. Erection sequence, if required.
5. Openings in precast concrete units (size and location).
6. Width of joints.
7. Grid lines related to joints.
8. Pertinent grid and structure lines.
9. Section marks.
10. Structural slab and floor elevations.
11. Elevation of top and/or bottom of precast concrete units.
12. Materials and structures adjacent to precast units in place at time of erection.
13. Location of connection inserts and other hardware for each unit.
14. Non-typical returns and blockouts.
15. Finished grade in relation to units.

Erection Plans

Erection plan drawings are required to show layout of precast floor and roof units. Plan drawings are especially useful for indicating beams, slabs or tees, columns, column covers, handrails, planters and coping. They are also used when elevations do not clearly define a specific area. Erection plans should include:

1. Pertinent grid lines in the vicinity of the precast units.
2. Size and orientation of any columns and beams or other precast units.
3. Precast supports and any variations.
4. Returns, openings and blockouts in structure.
5. Materials and structures adjacent to precast units in place at time of erection.
6. Anchor locations shown schematically for each unit mark.
7. Dimensions from centerline of anchor to sides or ends of unit.
8. Vertical location of plan by giving the floor line, roof level or top of precast elevation.
9. If the precast unit is to have a particular orientation or location, the identifying mark may be shown at one end rather than at the center. Placing the mark at one end of the member identifies its position on the job since the unit is marked in the same way on the framing plan.
10. True north and building north.
11. Estimated camber at erection of prestressed members.

Section Drawings

Sections are used to identify in detail the relationship of precast to precast, precast to structure and precast to grid. Section views and details are normally required on all plan and elevation drawings, and should include:

1. Precast units.
2. Finishes on all surfaces, when necessary.
3. Dimensions for: (1) unit size, (2) unit relationship to grid, including all clearances to structure.
4. Size of structural supports if they affect the location or function of a connection.
5. Architectural surroundings (without going into elaborate detail) if pertinent to attachment or finish so that the relationship of the precast member to the balance of the structure or structural framing can be seen clearly.
6. Elevations should be noted for beams, columns, column haunches, top of structural support, top and bottom of precast panel, and grade.

7. The clearances between precast units and columns, beams and slabs.

8. Grid lines running parallel to precast member. Note perpendicular grid or refer to elevation drawings.

9. Mark, type and location of the following hardware: (1) hardware cast into panels in the plant, (2) hardware in cast-in-place concrete or attached to steel frame, (3) erection hardware.

10. Dimension of joints.

11. Size, type and location for all flashing reglets and nailers, when necessary.

12. Location of all required welds showing size and length.

13. Size and location of all openings.

14. Any special erection sequence.

15. Installation sequence of interfacing materials, if necessary.

16. Temporary connections, if required, with instructions for erection sequence and adjustment procedure, especially in regard to possible deflection of horizontal supporting elements.

17. Erection restrictions posed by the structural materials, if required.

18. Notation of possible corrosion protection for details of lifting devices where such hardware is left in the units. If the handling devices interfere with any other trades, drawings should include removal instructions, as well as possible field repairs.

19. When specified by the contract documents, the method of sequence of erection, method of plumbing units and adjusting connections, and the loads and movements.

20. Mention of special protection of inserts or pockets against moisture collecting and freezing.

21. Use of permanent shims or temporary shims and removal, if required.

22. Size and elevation of structural frame which will receive precast units.

23. Instructions regarding the nature of the connections and the execution of the connections must be clearly indicated, for example, the location of the bolt in the slotted hole, the direction of the slotted hole, the tightness of the nut (torqued or finger tight), the size of the weld, whether it is a structural or tack weld; instructions regarding free movement of a sliding joint; instructions regarding the maximum size of shims and the adjustment due to the movements, tolerances and deflections.

24. The location of the bearing point in relation to the shims and tolerance for shim placement.

25. Information relating to grouting and drypacking such as materials and procedures, if required.

26. The type, location and orientation of all bearing pads.

Handling Detail

A handling detail conveys to the erector the handling procedures that have been selected by the precaster. It provides, in sequence, all procedures for a given unit or group of units for shipping, erection, and shoring, if required. It should be given to the erector in a timely manner, and should include:

1. Mark numbers and quantities.

2. All blocking points.

3. Which inserts (shown dimensionally) are used for each operation.

4. Location and type of all dunnage and/or padding required.

5. Hoisting diagram for rotating units.

6. Allowable angle of lifting lines, if critical to handling.

7. Method of lifting whenever unusual procedures must be followed.

8. Type and manufacturer of lifting anchors along with bolt penetration.

Anchor Plans

Anchor plans generally initiate the first direct contact between the precast erector and jobsite personnel. These are the drawings used to clearly locate hardware (either cast-in or pre-welded to the structure and/or foundations) for the purpose of connecting the precast units.

Ideally these drawings should be able to be used without the use of other reference drawings and therefore have their own basic sections and details. This way, the contractor who is settting this hardware or the erector's lay out person who is checking the layout ahead of erection need not be burdened with other drawings nor must they have an understanding of the structure beyond the specific location of this connecting hardware. Anchor plans should generally include:

1. Plan or elevation view of area to be represented as well as key plan to clearly show relationship of this area to the total structure. These plans or elevations

need to be clearly labeled (e.g., Second Floor CIP Plan, or West Elevation Along Grid 3).

2. True as well as building north.

3. Anchor locations to be dimensioned from nearest column or grid line not from edge of cast-in-place concrete.

4. Each column anchor bolt pattern should be clearly identified and detailed, showing relationship to grid lines as well as minimum projection of the bolts. If a pattern is non-symmetrical and is used in more than one orientation, there should be a separate detail for each orientation to avoid confusion and misplacement of the anchor bolts.

5. All depressions or raised portions of a support element required for the erecting of the precast. Location and size would be given dimensionally on the plan and can be clarified in a section or detail.

6. Type and size of anchor indicated in plan, and a section cut through each type of hardware and through different applications of each type.

7. In cast-in-place concrete framing applications, reinforcement which may be in the vicinity of a cast-in anchor should be shown.

8. Hardware, inserts or additional piers required for temporary bracing of precast units.

9. A bill-of-materials summarizing quantities of anchors required, with a brief description of each type, as well as identifying who supplies this hardware (i.e., contractor, steel supplier, or precaster).

APPENDIX D — WELDING

The strength of a welded connection depends on reliable workmanship and the compatibility of welding materials with the metals to be joined.[32,33]

Welding at temperatures below 32 deg. F (0 deg. C) may require preheating of materials to be joined because of the possibility of fractured welds. When welding in cold temperatures, preheating or special welding techniques should be used to prevent expansion of embedded steel that results in spalling of the adjacent concrete. When welding reinforcing bars or other structural components embedded in concrete, thermal expansion of the steel and transformation of moisture in the concrete to steam may induce spalling or cracking in the surrounding concrete. The expansion and distortion of the steel may also destroy the bond between the embedded component and the concrete. This is particularly the case when expansion of the heated metal is restrained by concrete.

The extent of cracking of concrete and distortion of the metal is dependent on the amount of heat generated during welding and the stiffness of the steel member. Heat may be reduced by the selection of low-heat welding rods of small diameter. The amount of weld metal should be controlled because large continuous welds can be detrimental to the unit and to the connection. Large or long welds should be made in stages (multiple passes), or as a series of intermittent welds, allowing the heat to dissipate. Wire brushing or grinding the weld area to ensure bare metal contact also reduces the heat required to successfully complete the weld. The person who actually performs the welding should also be warned of this problem and encouraged not to overweld the connection.

There are four primary reasons to utilize preheat for welding: (1) it slows the cooling rate in the weld metal and base metal, producing a more ductile metallurgical structure with greater resistance to cracking; (2) the slower cooling rate provides an opportunity for hydrogen that may be present to diffuse out harmlessly, reducing the potential for cracking; (3) it reduces the shrinkage stresses in the weld and adjacent base metal, which is especially important in highly restrained joints; and (4) it raises some steels above the temperature at which brittle fracture could occur in fabrication. Additionally, preheat can be used to help ensure specific mechanical properties, such as weld metal notch toughness.

Preheating carbon steel to a precise temperature generally is not required. Although it is important that the work be heated to a minimum temperature, it usually is acceptable to exceed that temperature by approximately 100 deg. F (40 deg. C).

When welded connections are required, welding should be performed by qualified welders in accordance with the project specifications and the erection drawings. The erection drawings should clearly specify type, size and length of weld, and, if critical, sequence, types of electrodes, minimum preheat and interpass temperatures, and location of welds. Temperature-indicating sticks are frequently used to gauge preheat temperature. This measurement should be made not closer than 3 in. (75 mm) of the weld on the base metal. Crayon marks should never be made directly on the weld because of possible contamination. When the insert plate temperature is below 32 deg. F (0 deg. C), the plate should be preheated to a temperature of at least 70 deg. F (21 deg. C) and this minimum temperature maintained throughout the entire welding process. The interpass temperature should be checked to verify that the minimum preheat temperature has been maintained just prior to initiating the arc of each pass. Three inches of the insert plate, in each direction around the weld, should be preheated to at least the minimum required temperature.

No unspecified weld should be added without approval of the precast engineer. Care should be taken to prevent overwelding which could cause additional forces in the connections due to restraint against volume changes. Overwelding a fillet increases the shrinkage force by the square of the weld size. Anything that reduces the amount of welding such as decreasing the leg size, reducing the weld length, or using intermittent welding techniques will minimize the shrinkage force and decrease distortion. Shrinkage and distortion can also be reduced by using fewer passes and using larger electrodes. Reducing the amount of welding should be done only with the approval of the precast engineer.

Welding procedures are contained in the ANSI/AWS D1.1[8] for structural steel and AWS D1.4[34] for reinforcing steel. AWS requirements include workmanship, technique, qualifications and inspection for acceptable welding. Welding parameters should be within the range recommended by the electrode manufacturer which is typically indicated on the package of electrodes.

Slag from each pass should be completely removed before depositing the next pass to avoid porosity and slag entrapment. Slag should be removed from all completed welds and the weld and adjacent base metal cleaned with a chipping hammer followed by brushing or other suitable means. Tightly adherent spatter remaining after the cleaning operation is acceptable. Accessible welds of corrosion-protected material (galvanized or painted) should be re-touched after welding. Zinc-rich paint should be brush or spray applied to a thickness of greater than 0.004 in. (100 μm) over the welded areas to replace the removed galvanizing or in conformance with ASTM A780.[35]

Structural welding which must be executed prior to the release of a unit from hoisting devices should be minimized, as hoisting time is crucial for economical erection. With welded connections, units may be temporarily shimmed while initial tack welding is accomplished. Provisions should be made to hold the unit safely in place while adjacent units are being placed and final adjustments are made. Before temporary supports are released, the full weld should be in place at every connection of the precast concrete unit. Temporary shims should be removed as soon as possible so that the connection system functions in the manner intended.

Surfaces to be welded, and surfaces adjacent to a weld, should be uniform, free from fins, tears, cracks and other discontinuities and free from loose or thick scale, slag, rust, moisture, grease, and other foreign material that would prevent proper welding or produce objectionable fumes. Mill scale that can withstand vigorous wire brushing, a thin rust-inhibitive coating, or antispatter compound may remain.

Welding should be avoided in enclosed or confined places and when hardware is galvanized, cadmium plated or stainless steel unless proper ventilation is provided.

When welding or cutting with a welding electrode, the ground lead should be attached directly to the base metal. Under no circumstances should the member be used as a conductor for the ground due to the possibility of internal arcing on principal reinforcement.

During welding, the erector should protect the unit and adjacent materials from damage resulting from weld spatter, smoke staining or excessive heat that can cause spalling. Non-combustible shields should be used to protect exposed concrete surfaces and glass in preglazed concrete panels during welding. Also, all work and materials of other trades in the area surrounding and below the welding area should be protected at all times. To minimize staining, all loose slag and debris should be removed immediately after the welding is complete.

A fire extinguisher of 5 BC or higher and in good operating condition should be located within reach of all burning and welding operations at all times. Welders working close to the perimeter and around floor openings should be protected against falls.

Galvanized Steel

One of the two following procedures should be used for welding galvanized steel:[36]

1. Removal of galvanizing on mating surfaces to be welded and use of standard welding procedures. The zinc coating should be removed at least 2 to 4 in. (50 to 100 mm) from either side of the intended weld zone and on both sides of the workpiece, if possible.

   When galvanized steel is welded, some of the zinc coating is volatilized on each side of the weld and, while a thin layer of zinc-iron alloy remains, there is a loss in corrosion resistance. In the case of zinc-rich painted steel, welding causes decomposition of the paint film which is burnt off for some distance each side of the weld. The width of the damaged zone will depend on the heat input and preheat.

   Removal of zinc, where weld is to be placed, is the most conservative approach in welding galvanized steel. Welding procedures will then be the same as for uncoated steel. Zinc can be removed by burning with a carbon arc or an acetylene torch while using an oxidizing flame, by shot blasting with portable equipment or grinding with silicone carbide abrasive discs.

2. Leave the galvanized coating and weld using galvanized base metal with the thickest coating anticipated and qualified by test in accordance with ANSI/AWS D1.1 or AWS D1.4. This Welding Procedure Specification (WPS) will permit welding over surfaces with coatings equal to or less than the coating used in qualification testing. In general, to avoid zinc penetration of the welds, the procedure should involve greater root openings in joints, electrodes with low silicon content and slower welding speeds.

   It is not uncommon to find small cracks in a fillet weld on galvanized steel, extending from the root toward the face of the bead. Whether cracking will occur depends on many factors such as the silicon content of the weld metal, the degree of penetration of the weld, the thickness of the base metal which influences restraint of the joint, and the coating weight of the zinc and the microstructure of the zinc coating which are both influenced by the composition of the base plate, particularly with respect to silicon content. Low-silicon or rutile (non-low-hydrogen) base electrodes with low-silicon content (0.2Si or lower) generally reduce cracking.

It is possible to obtain full penetration in the flat position on galvanized steel with the 1/16th in. (1.6 mm) root opening normally used on uncoated steel if the following conditions exist: (1) the angle the electrode makes with the plate is reduced from the usual 70 deg. to 30 deg.; (2) the electrode is moved backward and forward in a whipping motion in line with the joint.

In the case of vertical and overhead butt joints in galvanized steel, a root opening of 3/32 in (2 mm) is necessary to obtain full penetration. A root opening of 1/16 in. (1.6 mm) is sufficient to give full penetration in the horizontal position but the root pass tends to have a peaky or bulbous shape resulting in slag entrapment at the edges of the weld. This slag can only be removed by grinding. Slag entrapment can be prevented by increasing the root opening from 1/16 in. (1.6 mm) to 3/32 in. (2 mm).

When the root pass has been made, subsequent weld passes can be made with similar techniques to those used for uncoated steel because the weld beads are deposited largely on previously uncoated weld metal and only partially on the galvanized beveled edges. There is usually insufficient zinc present to affect the welding operation. If excessively thick zinc coatings are present, the oscillating motion of the electrode should be used to slightly reduce travel speed.

Welding of galvanized steel generally requires that the welder receive specialized training. In addition, qualifications of the welder and welding procedure using the thickest coating anticipated is strongly recommended.

The following points should be noted when welding galvanized steel:

1. The welding electrode EXX10 and EXX11 should be applied slower than usual with a whipping action which moves the electrode forward 1/8 to 5/16 in. (3 to 8 mm) along the seam in the direction of progression and then back into the molten pool. All volatilization of the galvanized coating should be complete before bead progresses. This will prevent zinc entrapment in the weld metal.

2. A short arc length is recommended for welding in all positions to give better control of the weld pool and to prevent either intermittent excess penetration or undercutting.

3. Slightly wider gaps, up to 3/32 in. (2 mm) are required in butt joints or a 15 deg. angle on the edge of a standing plate in order to give complete penetration. The gap also allows for the zinc and its gases to escape, and reduces cracking caused by restraint as the plate thickness increases.

4. Weaving and multiple weld beads should be avoided. Heat input into the joint should be kept to a minimum to avoid undue damage to the adjacent coating.

When welding galvanized steel, hydrogen-induced cracking of the heat-affected zone may occur in the base plate adjacent to the weld. The precautions necessary for the avoidance of cracking include such measures as reduction in cooling rate of the joint by the use of preheat or the use of large diameter electrodes at high currents. The hydrogen content of a weld can be increased when it is deposited on galvanized or zinc-rich primed steel. This extra hydrogen originates from the pickling process in galvanizing, or from the decomposition products of primers. It may be necessary to either remove primers from the vicinity of the joint before welding or use a higher preheating temperature than would be used on uncoated steel.

Zinc-rich paints are available which have been specially formulated such that it is not necessary to remove the coating from the weld path prior to welding. A letter should be obtained from the paint manufacturer stating it to be "weldable."

Porosity will occur in certain weld joint designs in galvanized steel, depending on coating thickness, due to volatilization of the zinc in the coating and entrapment of gas in the weld. The type of joint affects porosity formation since gases cannot readily escape from tee joints or from butt joints in thick materials. In the case of tee or butt joints, a vee edge preparation or provision of a gap between the plates facilitates the escape of gases, minimizing porosity. Pore formation is also influenced by the thickness of the galvanized coating relative to the base steel.

When a fillet weld in galvanized steel is large enough relative to plate thickness to fail by fatigue from the toe of the weld in the same manner as in uncoated steel, the presence of porosity in the weld does not reduce the fatigue strength of the joint. Where the dimensions of a weld are just large enough to cause fatigue failure from the toe in a sound weld, a weld containing porosity at the root may fail preferentially through the throat of the weld. Intergranular cracking of fillet welds containing porosity sometimes referred to as zinc penetrator cracking, may also affect the strength of welded joints and it is advisable to conduct procedural and welder tests on materials and samples.

If galvanized material must be welded, all weld areas should be given a coat of zinc-rich paint (95% zinc) or epoxy paint immediately after welding and chipping of slag to replace the removed galvanizing. Also, hardware should be properly cleaned by scouring with a stiff wire brush prior to application of protective treatment. All weld areas on primed, exposed or accessible steel anchorage devices should be painted with a rust inhibitive primer. Special care should be taken when welding in the vicinity of bearing pads to avoid damaging the pads.

Stainless Steel

When welding stainless steel plates to other stainless steel or to low carbon steel, the general procedure for welding low carbon steels should be followed, taking into account the stainless steel characteristics that differ, such as higher thermal expansion and lower thermal conductivity.[37] Welding of stainless steel should be done by qualified welders familiar with the welding requirements of these alloys. Austenitic stainless steels are best welded without preheat except to reduce shrinkage stresses on sections over 1-1/4 in. (32 mm) thick or on restrained joints. No preheat, interpass temperature below 350 deg. F (177 deg. C), or a stringer-bead technique reduce the time the heat-affected zone is in the sensitizing range [800 to 1400 deg. F (427 to 760 deg. C)], thereby reducing the amount of carbide precipitation and warpage. Preheating will be necessary when carbon steel is to be welded or temperature of components to be welded is less than 32 deg. F (0 deg. C).

Temperature indicating sticks used on austenitic stainless steels should have a certified analysis which meets the following specified maximum amounts of detrimental contaminants:

1. The inorganic halogen content less than 200 ppm by weight.
2. The halogen (inorganic and organic) content less than 1 percent by weight.
3. Sulfur content less than 1 percent by weight as measured in accordance with ASTM D129.[38]
4. Total content of low-melting-point metal (lead, bismuth, zinc, mercury, antimony, and tin) less than 200 ppm by weight and no individual metal content greater than 50 ppm by weight.

A short arc should be used but the coating kept from touching the puddle. Flat beads with good wash-in promote easy slag removal in deep grooves. Each crater should be filled before breaking the arc, to avoid crater cracks. The slag should be thoroughly cleaned from the finish of the bead before starting another electrode, and the complete weld cleaned before starting the next pass. On deep-groove butt joints, the root pass should penetrate only enough to fuse to both plates and seal the opening. More penetration may cause cracks.

Stainless steel welds generally are not as penetrating as those in carbon steels. And the weld metal is a bit more sluggish. This means that the weld joints in stainless have to be a bit more open to obtain full penetration and fusion.

For vertical and overhead positions, electrodes should be 5/32 in. (4 mm) in diameter or less. For vertical welds on 3/16 in. (5 mm) thick plate or less, the weld should be vertically down using small beads; otherwise, on thicker plates, the triangular weave technique should be used, welding vertically up.

Because of the relatively high coefficient of thermal expansion and lower thermal conductivity of austenitic stainless steel, precautions should be taken to avoid weld bead cracking, minimize distortion of steel and avoid cracking of concrete. The following procedures may be used to minimize these problems: lower weld current consistent with sufficient penetration to reduce the heat input to the work, skip-weld techniques to minimize heat concentration, use of other cooling techniques to dissipate heat, tack welding to hold the parts in alignment during welding and varying sequence of welding. When stainless steel connection plates are used, edges should be kept free from adjacent concrete to allow expansion during welding without spalling the concrete.

Surfaces to be welded should be sanded smooth, not ground, and all blue heat tint removed. Joints should be clean and dry. Moisture should be removed by heating or by blowing with dry air. Prepared surfaces on which weld metal is to be deposited should be free from organic contaminants, surface oxides or material discontinuities (e.g., tears, laminations) that would affect the strength of the weldment. In the absence of specific contract documents for stainless steel, the surface discontinuity limitations of AWS D1.1 may be used as a guide.

In joining austenitic stainless steels to carbon steels or low-alloy steels, a stainless steel welding rod that is sufficiently high in total alloy content, such as Type 309 or 309L or other highly alloyed electrodes, should be used. Buttering may be an unnecessary and costly procedure. However, when joining austenitic stainless steel and carbon steel, it may be a good practice to "butter" the carbon steel surface with a layer of Type 309 or other suitable stainless steel weld metal, such as Type 312, prior to actually joining it to the stainless steel. The portion of the joint where difficulties are most likely to occur is buttered while there is little restraint on the weld metal. Following the deposition and inspection of the buttered layer or layers, the joint between the stainless steel member and the buttered layer will be a conventional stainless steel to stainless steel joint. The welding electrode in this case can be the type normally used to weld the stainless steel member of the joint (i.e., Type 308 if the stainless steel member is Type 304).

Carbon steels with less than 0.20% carbon can normally be welded with austenitic fillers without preheat, but when the carbon is greater than 0.30% temperature control is necessary. As alloy content increases, i.e., in the case of low-alloy steels, preheat control is usually essential.

The deposition of carbon steel or low-alloy steel weld metal on stainless steel can result in hard, brittle weld deposits which frequently crack when deposited and which would be likely to fail in service. Weld procedure qualifications should be conducted on all proposed combinations of base metal and electrodes to determine the extent to which this is a problem.

When welds are made using stick electrodes (SMAW), some spatter is normal. However, it is easily removed by light grinding (aluminum oxide) or wire brushing. Spatter-resisting compounds applied before welding reduces this annoyance. Cleaning time is reduced or eliminated when welding is done using the inert gas process.

Welds and the surrounding area on stainless steel should be cleaned of weld spatter, flux, or scale to avoid impairment of corrosion resistance. Most welding flux or slag can be chipped off, but it is better to grind the flux off using clean grinding wheels. Sandblasting with clean silica sand is also an effective and economical way to remove slag. Stainless steel brushes and clean grinding wheels (wheels not contaminated with carbon steel particles) should be used to clean stainless steel welds. A small piece of carbon steel, scale, copper, or foreign material lodged on stainless steel may be sufficient to destroy passivity at the point of contact and cause rust staining and increased corrosion.

Reinforcement

When welding of reinforcement is used to make a connection, weldability characteristics of the reinforcement should be considered. If required, minimum preheating and interpass temperatures for welding of reinforcing bars should consider the highest carbon equivalent number of the base metal. The weldability of steel established by its chemical analysis determines the applicable welding procedures and sets preheat and interpass temperature requirements. Striking an arc outside of the weld area of the reinforcing bar should be avoided.

One practical method of preheating reinforcing bars and insert plates is to use an oscillating torch while measuring the preheat with a temperature level or heat sensitive chalk. By using the torch, a localized preheated area can be obtained simplifying the welding process. The entire plate or bar does not need to be heated. Heat sensitive chalk is designed to change color or melt when the desired temperature is reached.

If preheating is required, the reinforcing bar should be heated until the cross-section of the bar is at or above the minimum temperature for at least six in. (150 mm) on either side of the joint.

Surfaces of reinforcing bars to be welded and surfaces adjacent to a weld should be free from loose or thick scale, slag, rust, moisture, grease, epoxy coating, or other foreign material that would prevent proper welding or produce objectionable fumes. Mill scale that withstands vigorous wire brushing, a thin rust inhibitive coating, or anti-spatter compound, may remain.

The ends of reinforcing bars in direct butt joints should be shaped to form the weld groove by oxygen cutting, air carbon arc cutting, sawing, or other mechanical means. Bars for direct butt joints that have sheared ends should be trimmed back beyond the area deformed by shearing.

Welding galvanized reinforcing bars, without prior removal of the coating, should be performed in accordance with AWS D1.4. Welding of galvanized metal may also be done after removing all coating from within 2 in. (50 mm) of the weld joint. The galvanized coating should be removed with oxyfuel gas flame, abrasive shot blasting or other suitable means.

When welding or preheating epoxy coated base metal, the epoxy coating should be removed from the surfaces to be heated. After welding, suitable protective coating (zinc-rich or epoxy paint) should be applied to the finished joint to restore the corrosion resistance properties of the coated bars.

When reinforcing steel is welded to structural steel members, the project specification for structural steel should apply. When joining different grades of steel, the filler metal should be selected for the lower strength base metal.

Welding Electrodes

Electrodes for shielded metal arc welding (SMAW) should conform to the requirements of ANSI/AWS D1.1, Section 4 or AWS D1.4, Section 5 (AWS A5.1 or A5.5).[39] All welding electrodes should be of a type suitable for the chemistry of the steel being welded.

The electrodes and the shielding for gas metal arc welding (GMAW) or flux-cored arc welding (FCAW), for producing weld metal with minimum specified yield strengths of 60 ksi (415 MPa) or less, should conform to the requirements of AWS A5.20 or A5.29.[39]

Weld metal having a minimum specified yield strength greater than 60 ksi (415 MPa) should conform to the following requirements.

1. The electrodes and shielding for gas metal arc welding for producing weld metal with a minimum specified yield strength greater than 60 ksi (415 MPa) should conform with AWS A5.28.[39]

2. The electrodes and shielding gas for flux-cored arc welding for producing weld metal with a minimum

specified yield strength greater than 60 ksi (415 MPa) should conform to AWS A5.20 and A5.29.

Electrodes similar to those used for arc welding uncoated steel can be used for welding galvanized steel, along with the same electrode size and current. The major difference when manual arc welding galvanized steel compared to uncoated steel arises from the need for higher heat input to remove the zinc from the weld pool and lower welding speed to burn off as much as possible of the zinc coating in front of the weld pool.

All low hydrogen electrodes conforming to AWS A5.1 should be purchased in hermetically sealed containers or should be dried for at least 2 hrs. at a temperature between 500 deg. F (260 deg. C) and 800 deg. F (430 deg. C) before they are used, if the container is damaged.

Storage temperatures for E6010 and E6011 (high cellulosic type) electrodes are not critical but temperatures exceeding 125 deg. F (52 deg. C) should be avoided since these electrodes depend on a high percentage of moisture (6 to 7%) for favorable arc characteristics; over-baked or dried-out cellulosic electrodes result in low arc force, low penetration, incomplete slag coverage and porosity.

Of the various standard electrode covering types, low-hydrogen coverings are the most resistant to hydrogen-induced cracking. These types of inorganic electrodes are produced with a minimum moisture content in their coatings and are given a high-temperature bake before packaging in hermetically sealed containers designed to provide adequate protection from exposure under normal conditions.

Electrodes that have been in physical contact with water, grease or oil should not be used, and it is not recommended to re-bake those electrodes. These electrodes should be scrapped.

Condensation on covered electrodes will occur when cans or cartons are opened prematurely in a relatively warm area after removal from storage in a relatively cool area. Containers should be allowed to warm up to ambient temperature before opening in the welding location. Cartons will require more time than cans to equalize temperature due to the insulating effect of the cardboard.

Welding with "moist" low hydrogen electrodes leads to hydrogen-induced cracking, increased arc voltage, spatter loss, undercutting, and poor slag removal, in addition to porosity, underbead cracking and general rough welds. Therefore, all low hydrogen and stainless steel shielded metal arc electrode coverings should be protected from moisture pickup. Electrodes packaged in hermetically sealed containers can be stored for several months without deterioration. However, electrodes should be dried as specified above prior to use if the hermetically-sealed containers are damaged, improperly stored, or for any reason the electrodes are exposed to high moisture conditions. Immediately after removal from hermetically sealed containers or from drying ovens, electrodes should be stored in holding ovens and held at a temperature of 250 deg. F (120 deg. C). E70 series electrodes that are not used within 4 hrs., E80 series within 2 hrs., E90 series electrodes within 1 hr., or E100 series within 1/2 hr., after removal from hermetically sealed containers or from a drying or storage oven, should be redried in a drying oven before use. The manufacturers of electrodes may warrant longer exposure periods based on tests. Electrodes should be dried only once for any reason. Electrodes which have been wet should not be used.

When drying electrodes, stack them in a maximum of three layers. The following method of reconditioning works best:

- Load electrodes into a preheated oven that does not exceed 200 deg. F (93 deg. C).
- Increase to recommended temperature for conditioning (Table D-1) and hold for about one hour. When oven temperature exceeds 500 deg. F (260 deg. C), the heating rate should not exceed 300 deg. F (149 deg. C) per hour.
- Cool the oven to 150 to 200 deg. F (66 to 93 deg. C) before unloading electrodes.
- Store as recommended in Table D-1.

The bond between the covering of the electrode and the core wire can loosen if electrodes are loaded or unloaded at, or close to, the baking temperature or if the baking cycle is repeated indefinitely.

Only low hydrogen welding rods (EXX-X5, 6, or 8) should be kept in an oven once removed from their airtight container. Although they must not get wet, the coating of other rods (60-11) will be damaged if heated.

All stainless steel SMAW electrode coverings are of the low-hydrogen type and must be protected from moisture. Electrodes should be purchased in hermetically sealed containers, which can be stored for several months without deterioration.

The hazard of moisture pick-up is not as great with cored wires as it is with covered electrodes. The low-hydrogen fill ingredients of cored wires are to a considerable extent protected from the air by the tight wrapping of the steel sheath. Nevertheless, it is possible to damage the low-hydrogen character of cored wires by excessive exposure to moisture-laden air. This is most likely to occur by overnight exposure to high humidity. Returning the partially used coil or spool of wire to its plastic bag when the wire will not be used for a period of time (e.g., 8 hours or more) is a simple but effective means of preventing excessive exposure.

Electrode Type	Storage of Contents of Open Cartons	Reconditioning
Mild Steel – 6010, 6011	Dry at room temperature	Not recommended
Mild Steel – 6012, 6013, 7014, 7024	100 – 175° F (38 – 80° C)	250 – 300° F (121 – 149° C)
Mild Steel – Low Hydrogen, Low Alloy EXX15, EXX16, EXX18	215 – 450° F (102 – 233° C)	700 – 800° F (371 – 427° C)
Stainless Steel Electrodes		
AC/DC (AWS-16), DC Lime (AWS-15)	215 – 260° F (102 – 127° C)	400 – 600° F (205 – 316° C)
Sterling (AWS-17)	215 – 260° F (102 – 127° C)	350 – 425° F (177 – 219° C)

Table D-1 — Recommendations for Storing and Reconditioning Electrodes.

When joining different grades of steel, the electrode should be selected for the lower strength base metal.

If the weld joint in a low carbon, unalloyed steel cannot be completely cleaned of scale, rust, paint, moisture or dirt, an E6010 or 11 electrode should be used, except for steels having yield strengths in excess of 70 ksi. The E6010 and E6011 electrodes are sources of large amounts of hydrogen and in high strength steels hydrogen-induced cracks may occur.

When welding stainless steel, it is necessary to select a welding rod that provides weld filler metal having corrosion resistance properties as nearly identical to the base metal as possible, or better. The characteristics of the weld metal are primarily dependent on the alloy content of the filler metal and to a lesser extent on the degree to which the molten weld metal is protected from the environment. This protection is provided by the shielding gases used in certain welding processes or by the action of chemical fluxes applied to welding rods. The greater amount of chromium and nickel in certain alloys, Type 308 for example, is useful for welding Types 302 and 304 base metals and hence is standard for all the lower chromium-nickel base metals. While the same principle applies to Type 316, in that the minimum chromium is higher in the weld metal than the base metal, the designation of the filler metal is the same.

The maximum diameter of electrodes should be:

1. 5/16 in. (8.0 mm) for all welds made in the flat position, except root passes,

2. 1/4 in. (6.4 mm) for horizontal fillet welds,

3. 1/4 in. (6.4 mm) for root passes of fillet welds made in the flat position and groove welds made in the flat position with backing and with a root opening of 1/4 in. (6.4 mm) or more,

4. 5/32 in. (4.0 mm) for welds made with low hydrogen electrodes in the vertical and overhead positions,

5. 3/16 in. (4.8 mm) for root passes of groove welds and for all other welds not included above.

DC electrodes (EXXX-15) operate on DC only, have good penetration, produce fillets with a slightly convex profile, and are recommended for:

1. Vertical and overhead welding and all-position applications such as pipe. The slag has a fast freeze characteristic.

2. Root passes on heavy plate. The larger throat section of the convex bead helps prevent cracking.

3. Austenitic stainless welds that cannot contain any ferrite.

AC-DC electrodes (EXXX-16) are always used on DC when this type of power is available. The fillet profile is flat to slightly concave, the weld surface is smoother and the penetration is less than with EXXX-15 (DC only) electrodes. The larger amount of slag requires more care to avoid slag inclusions. These electrodes are recommended for horizontal fillets and for all flat position welding. EXXX-16 electrodes are also used in all positions by skilled welders.

If Type 304 stainless steels are thick enough to require multiple pass welds, carbide precipitation, which can cause intergranular corrosion, is likely to occur. Extra low-carbon plate (less than 0.03%) 304L, and electrodes such as 308L be used to minimize carbide precipitation.

One of the most common dissimilar metal weld combinations is Type 304 stainless to a low carbon or mild steel. High alloy filler metal such as Type 309 or Type 312 should be used to make this weld to maintain crack or fissure resistance.

Electrodes for welding copper silicon (silicon bronze) and copper aluminum (aluminum bronze) should conform to AWS A5.6 or A5.7.[39]

WELDING PROCEDURE SPECIFICATION (WPS) YES_X_
PREQUALIFIED_X_ QUALIFIED BY TESTING_____
OR PROCEDURE QUALIFICATION RECORDS (PQR) YES____

Identification #____
Revision_NEW_ Date_____ By____
Authorization by_____ Date____
Type - Manual_X_ Semi-Automatic____
Machine____ Automatic____

Company Name____
Welding Process(es)_SMAW_
Supporting PQR No.(s)_N/A_

JOINT DESIGN USED
Type: Single_X_ Double Welded _N/A_
Backing: Yes_N/A_ No_N/A_

Backing MaterialANY AWS TABLE 3.1 GRP1MATERIALS
Root Opening_1/4"_Root Face Dimension_1/32"_
Groove Angle_45 Degrees_Radius (J-U) _N/A_
Back Gouging: Yes____No_X_ Method____

BASE METAL
Material Spec_ALL AWS TABLE 3.1GRP1MATERIALS_
Type or Grade_ALL AWS TABLE 3.1GRP1 MATERIALS_
Thickness: Groove_1/8" -UNLIMITED_Fillet____
Diameter (Pipe)_OVER 24" DIAMETER_

FILLER METALS
AWS Specification_SFA 5.1, 5.5_
AWS Classification_E7018_

SHIELDING
Flux____ Gas_NONE_
Composition____
Electrode-Flux (class) Flow Rate____
____ Gas Cup Size____

POSITION
Position of Groove_ALL_ Fillet_N/A_
Vertical Progression: Up_X_ Down____

ELECTRICAL CHARACTERISTICS
Transfer Mode (GMAW):
Short Circuiting____Globular____Spray____
Current: AC____DCEP_X_DCEN____Pulsed____
Other ____
Tungsten electrode (GTAW):
Size____
Type____

TECHNIQUE
String or Weave Bead_WEAVE OR STRING BEAD_
Multi-pass or Single Pass (per side) _Multi or Single_
Number of Electrodes_SINGLE_
Electrode Spacing: Longitudinal____
Lateral____
Angle____
Contact Tube to Work Distance____
Peening ____
Interpass Cleaning _Wire brush/Grind/Chip_

PREHEAT
Preheat Temp Min._If below 32 degrees F, preheat to 70 degrees F, and maintain during welding_
Interpass Temp. Min._N/A_ Max._N/A_

POSTWELD HEAT TREATMENT
Temp. _NONE_
Time_NONE_

WELDING PROCEDURE

Pass or Weld Layer(s)	Process	Filler Metals		Current		Volts	Travel Speed	Joint Details
		Class	Diam.	Type & Polarity	Amps or Wire Feed Speed			All Vgrooves with backing per figure 3.4 of AWS D1.1-98.
ALL LAYERS	SMAW SMAW SMAW	E7018 E7018 E7018	3/32" 1/8" 5/32"	DCEP DCEP DCEP	70-100 115-165 150-220		AS REQ. AS REQ. AS REQ.	

AWS D1.1 PREQUALIFED
JOINT DETAILS
(SMAW)

SEE NOTES ON PAGE 142

Single-bevel-groove weld (4)
Butt joint (B)

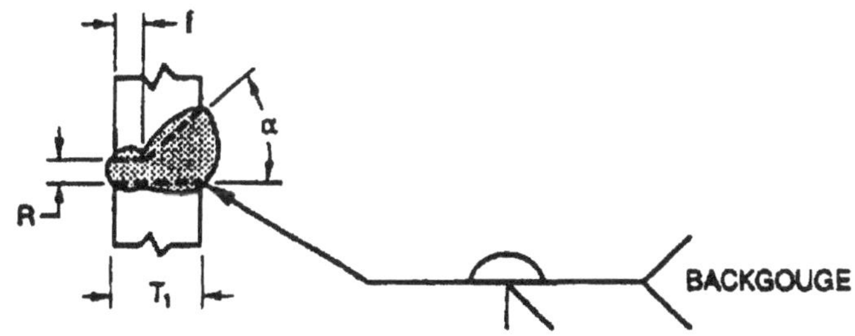

Welding Process	Joint Designation	Base Metal Thickness (U = unlimited) T_1	T_2	Groove Preparation: Root Opening, Root Face, Groove Angle	Tolerances: As Detailed	Tolerances: As Fit-Up	Permitted Welding Positions		Notes
SMAW	B-U4b	U	—	R = 0 to 1/8 f = 0 to 1/8 α = 45°	+1/16, −0 +1/16, −0 +10°, −0°	+1/16, −1/8 Not limited 10°, −5°	All		Br, C, D, N
							All		A, Br, C, N

Single-bevel-groove weld (4)
T-joint (T)
Corner joint (C)

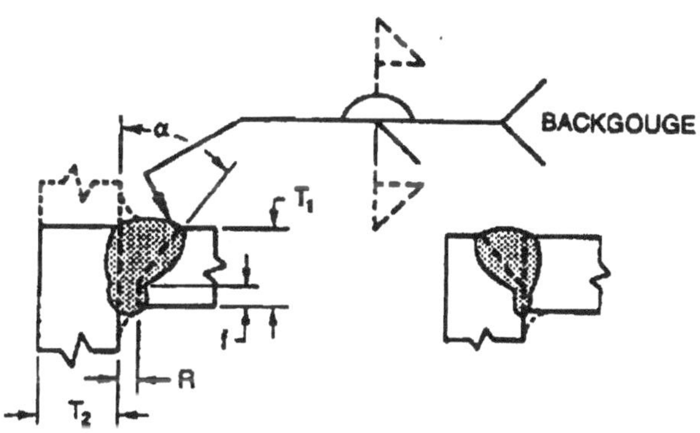

Welding Process	Joint Designation	Base Metal Thickness (U = unlimited) T_1	T_2	Groove Preparation: Root Opening, Root Face, Groove Angle	Tolerances: As Detailed	Tolerances: As Fit-Up	Permitted Welding Positions		Notes
SMAW	TC-U4b	U	U	R = 0 to 1/8 f = 0 to 1/8 α = 45°	+1/16, −0 +1/16, −0 +10°, −0°	+1/16, −1/8 Not limited 10°, −5°	All		C, D, J, N, V
							All		A, C, J, N, V

AWS D1.1 PREQUALIFED JOINT DETAILS (SMAW)

SEE NOTES ON PAGE 142

Single-bevel-groove weld (4)
T-joint (T)
Corner joint (C)

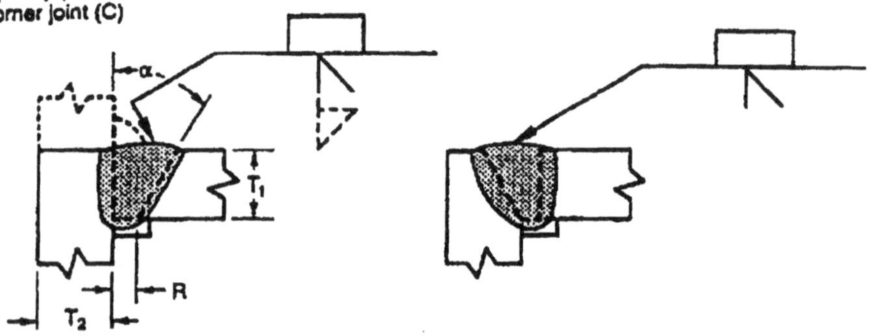

Tolerances	
As Detailed	As Fit-Up
R = +1/16, −0	+1/4, −1/16
α = +10°, −0°	+10°, −5°

Welding Process	Joint Designation	Base Metal Thickness (U = unlimited)		Groove Preparation		Permitted Welding Positions		Notes
		T_1	T_2	Root Opening	Groove Angle			
SMAW	TC-U4a	U	U	R = 1/4	α = 45°	All		D, J, N, V
				R = 3/8	α = 30°	F, V, OH		D, J, N, V

Single-bevel-groove weld (4)
Butt joint (B)

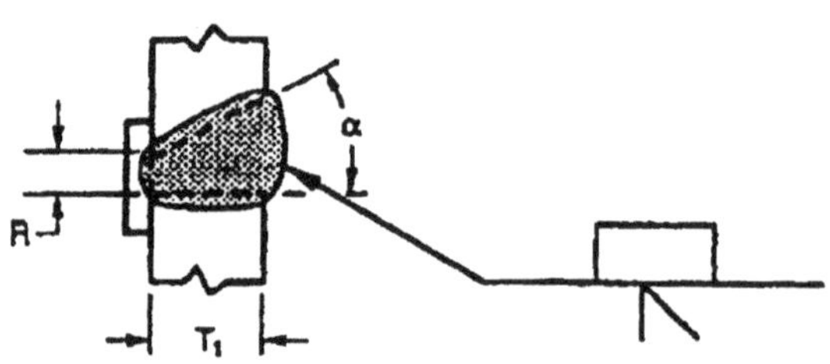

Tolerances	
As Detailed	As Fit-Up
R = +1/16, −0	+1/4, −1/16
α = +10°, −0°	+10°, −5°

Welding Process	Joint Designation	Base Metal Thickness (U = unlimited)		Groove Preparation		Permitted Welding Positions		Notes
		T_1	T_2	Root Opening	Groove Angle			
SMAW	B-U4a	U	—	R = 1/4	α = 45°	All		Br, D, N
				R = 3/8	α = 30°	All		Br, D, N

AWS D1.1 PREQUALIFED JOINT DETAILS (SMAW)

SEE NOTES ON PAGE 142

Double-V-groove weld (3)
Butt joint (B)

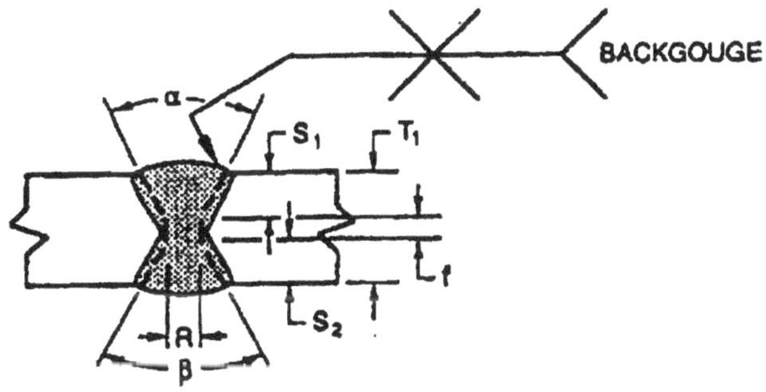

Welding Process	Joint Designation	Base Metal Thickness (U = unlimited)		Groove Preparation			Permitted Welding Positions		Notes
		T_1	T_2	Root Opening Root Face Groove Angle	Tolerances: As Detailed	Tolerances: As Fit-Up			
SMAW	B-U3b	U	—	R = 0 to 1/8 f = 0 to 1/8 α = β = 60°	+1/16, −0 +1/16, −0 +10°, −0°	+1/16, −1/8 Not limited +10°, −5°	All		C, D, M, N
							All		A, C, M, N

Double-bevel-groove weld (5)
Butt joint (B)

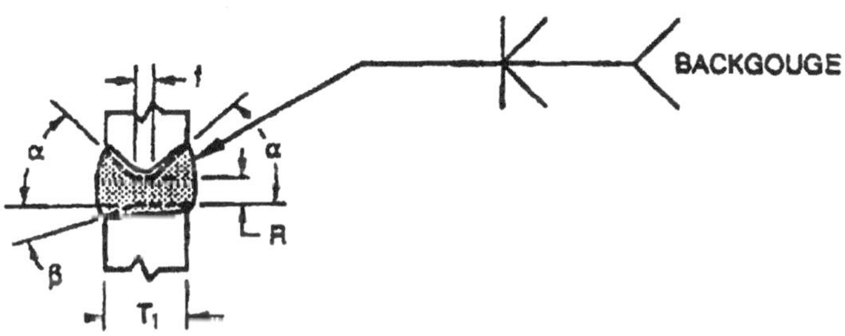

Welding Process	Joint Designation	Base Metal Thickness (U = unlimited)		Groove Preparation			Permitted Welding Positions		Notes
		T_1	T_2	Root Opening Root Face Groove Angle	Tolerances: As Detailed	Tolerances: As Fit-Up			
SMAW	B-U5a	U	—	R = 0 to 1/8 f = 0 to 1/8 α = 45° β = 0° to 15°	+1/16, −0 +1/16, −0 $\alpha + \beta^{+10°}_{-0°}$	+1/16, −1/8 Not limited $\alpha + \beta^{+10°}_{-5°}$	All		Br, C, D, M, N

AWS D1.1 PREQUALIFED
JOINT DETAILS
(SMAW)

SEE NOTES ON PAGE 142

Single-V-groove weld (2)
Butt joint (B)

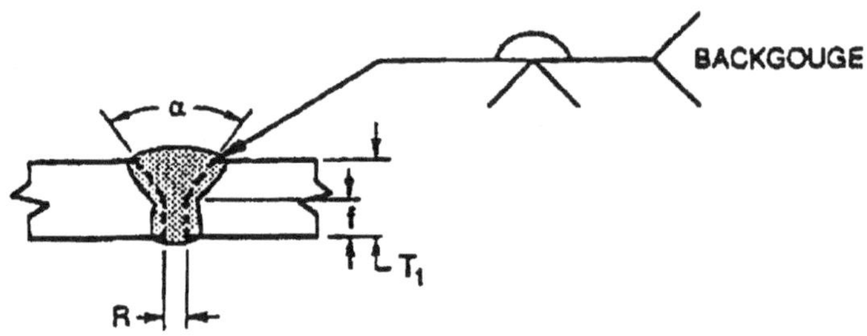

Welding Process	Joint Designation	Base Metal Thickness (U = unlimited)		Groove Preparation			Permitted Welding Positions		Notes
				Root Opening Root Face Groove Angle	Tolerances				
		T_1	T_2		As Detailed	As Fit-Up			
SMAW	B-U2	U	—	R = 0 to 1/8 f = 0 to 1/8 α = 60°	+1/16, −0 +1/16, −0 +10°, −0°	+1/16, −1/8 Not limited +10°, −5°	All		C, D, N

Single-V-groove weld (2)
Butt joint (B)

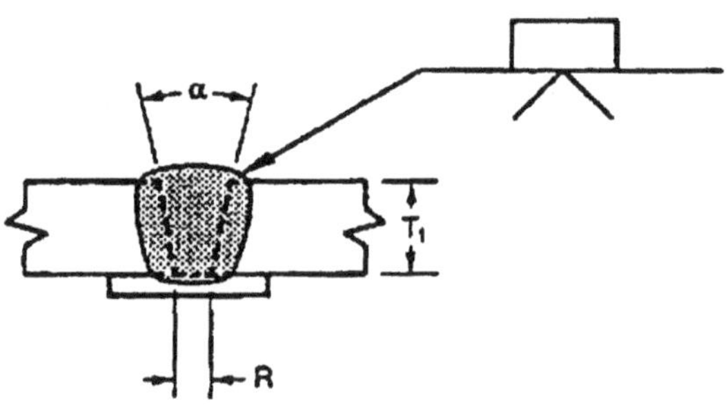

Tolerances	
As Detailed	As Fit-Up
R = +1/16, −0	+1/4, −1/16
α = +10°, −0°	+10°, −5°

Welding Process	Joint Designation	Base Metal Thickness (U = unlimited)		Groove Preparation		Permitted Welding Positions		Notes
		T_1	T_2	Root Opening	Groove Angle			
SMAW	B-U2a	U	—	R = 1/4	α = 45°	All	—	D, N
				R = 3/8	α = 30°	F, V, OH	—	D, N
				R = 1/2	α = 20°	F, V, OH	—	D, N

AWS D1.1 PREQUALIFED JOINT DETAILS (SMAW)

SEE NOTES ON PAGE 142

Double-bevel-groove weld (5)
T-joint (T)
Corner joint (C)

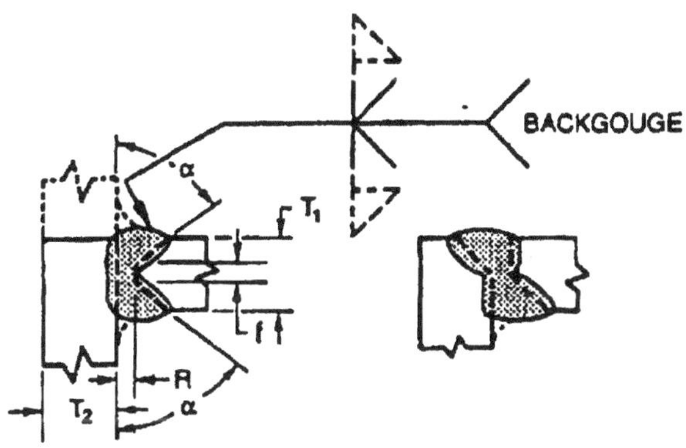

Welding Process	Joint Designation	Base Metal Thickness (U = unlimited) T_1	T_2	Groove Preparation: Root Opening, Root Face, Groove Angle	Tolerances: As Detailed	Tolerances: As Fit-Up	Permitted Welding Positions		Notes
SMAW	TC-U5b	U	U	R = 0 to 1/8 f = 0 to 1/8 α = 45°	+1/16, −0 +1/16, −0 + 10°, −0°	+1/16, −1/8 Not limited +10°, −5°	All		C, D, J, M, N, V
							All		A, C, J, M, N, V

Single-V-groove weld (2)
Corner joint (C)

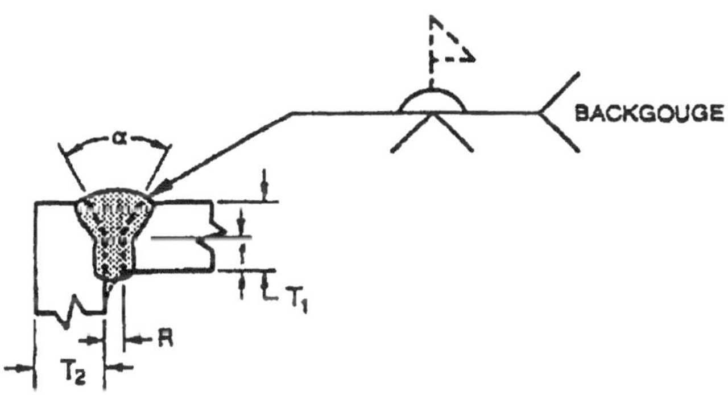

Welding Process	Joint Designation	Base Metal Thickness (U = unlimited) T_1	T_2	Groove Preparation: Root Opening, Root Face, Groove Angle	Tolerances: As Detailed	Tolerances: As Fit-Up	Permitted Welding Positions		Notes
SMAW	C-U2	U	U	R = 0 to 1/8 f = 0 to 1/8 α = 60°	+1/16, −0 +1/16, −0 + 10°, −0°	+1/16, −1/8 Not limited +10°, −5°	All		C, D, J, N

AWS D1.1 PREQUALIFED JOINT DETAILS (SMAW)

Single-V-groove weld (2)
Corner joint (C)

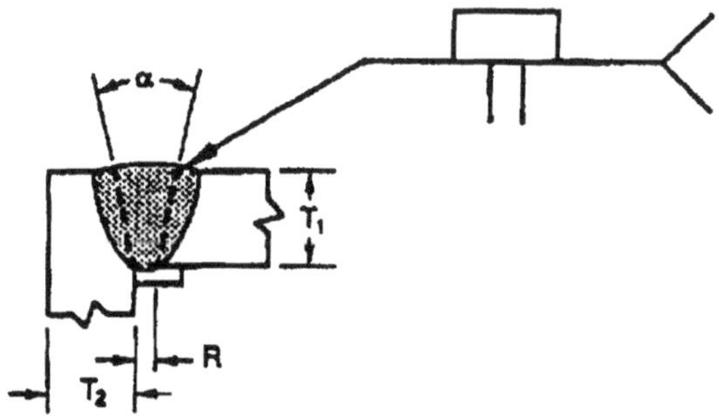

Tolerances	
As Detailed	As Fit-Up
R = +1/16, −0	+1/4, −1/16
α = +10°, −0°	+10°, −5°

Welding Process	Joint Designation	Base Metal Thickness (U = unlimited)		Groove Preparation		Permitted Welding Positions		Notes
		T_1	T_2	Root Opening	Groove Angle			
SMAW	C-U2a	U	U	R = 1/4	α = 45°	All		D, N
				R = 3/8	α = 30°	F, V, OH		D, N
				R = 1/2	α = 20°	F, V, OH		D, N

Notes:

A: Not prequalified for gas metal arc welding using short circuiting transfer nor GTAW.

B_r: Cyclic load application limits these joints to the horizontal welding position.

C: Backgouge root to sound metal before welding second side.

D: SMAW detailed joints may be used for prequalified GMAW (except GMAW-S) and FCAW.

J: If fillet welds are used in statically loaded structures to reinforce groove welds in corner and T-joints, these shall be equal to 1/4 T_1, but need not exceed 3/8 in. Groove welds in corner and T-joints of cyclically loaded structures shall be reinforced with fillet welds equal to 1/4 T_1, but not more than 3/8 in.

M: Double-groove welds may have grooves of unequal depth, but the depth of the shallower groove shall be no less than one-fourth of the thickness of the thinner part joined.

N: The orientation of the two members in the joints may vary from 135° to 180° for butt joints, or 45° to 135° for corner joints, or 45° to 90° for T-joints.

V: For corner joints, the outside groove preparation may be in either or both members, provided the basic groove configuration is not changed and adequate edge distance is maintained to support the welding operations without excessive edge melting.

Minimum Prequalified Partial Joint Penetration Weld Sizes

Base Metal Thickness of Thicker Part Joined	Minimum Weld Size*	
in. (mm)	in.	mm
1/8 (3.2) to 3/16 (4.8) incl.	1/16	2
Over 3/16 (4.8) to 1/4 (6.4) incl.	1/8	3
Over 1/4 (6.4) to 1/2 (12.7) incl.	3/16	5
Over 1/2 (12.7) to 3/4 (19.0) incl.	1/4	6
Over 3/4 (19.0) to 1-1/2 (38.1) incl.	5/16	8
Over 1-1/2 (38.1) to 2-1/4 (57.1) incl.	3/8	10
Over 2-1/4 (57.1) to 6 (152) incl.	1/2	13
Over 6 (152)	5/8	16

* Except the weld size need not exceed the thickness of the thinner part.

Welding positions

F — flat
H — horizontal
V — vertical
OH — overhead

Dimensions

R = Root Opening
α, β = Groove Angles
f = Root Face
r = J- or U-Groove Radius
S, S_1, S_2 = PJP Groove Weld Depth of Groove
E, E_1, E_2 = PJP Groove Weld Sizes corresponding to S, S_1, S_2, respectively

APPENDIX E — REPAIR PROCEDURES

A nominal amount of necessary product repair is to be expected. Repair work requires expert craftsmanship and careful selection and mixing of materials, if the end result is to be structurally sound, durable and aesthetically pleasing. Repairs will be acceptable if the structural adequacy, serviceability and the appearance of the product are not impaired. Excessive variation in color or texture of repairs compared to surrounding surfaces may result in non-acceptance of architectural precast units until the variation is minimized.

A written repair procedure prepared by the precaster should clearly define when to consult with the precast engineer to determine an appropriate repair. Major repairs should not be attempted until an engineering evaluation is made to determine whether or not the unit will be structurally sound. When repairable, the procedure should be approved by the precast engineer.

The selection of techniques or materials for the repair will depend on the following:

1. Extent of damage.
2. Function of the product.
3. Availability of equipment and skilled manpower.
4. Economic considerations.
5. Requirement for speed of repair.
6. The importance of appearance.

It is important that all repairs be performed in advance of final cleaning and joint sealing. Any repair work in joints should be fully cured, clean, and dry prior to caulking.

All repairs and remedial work should be documented and kept in job record files.

Repairs should be performed only when conditions exist which ensure that the repaired area will conform to the surrounding work with respect to appearance, structural adequacy and durability. Slight color variations may be expected between the repaired area and the original surface due to the different age and curing conditions of the repair. Time (several weeks or possibly months) will normally tend to blend the repair with the surrounding area so that it will become less noticeable.

Color differential between units can sometimes be improved by chemical treatment. Bleaches, acid, acid/water combinations using a pressure gun or an acrylic or silane sealer with toners can all be used in chemical treatment. Discoloration of concrete can sometimes be corrected by bleaching darker areas and using a pigmented sealer to blend the area with the surrounding concrete. Lighter-colored areas can also be treated with a pigmented sealer. Bleaches (oxidizing agents) which may be tried include a commercial sodium or potassium hypochlorite solution (for example, liquid household bleach); a 10-percent solution of phosphoric or oxalic acid; a 20- to 30-percent water solution of diammonium citrate; one part of sodium citrate in six parts of lukewarm water mixed thoroughly with seven parts of lime-free glycerol (glycerine); or one part of ammonium or sodium citrate in six parts of lukewarm water. These solutions should be applied to the surface and to a blanket of flannel, linen or cheese cloth which is placed over the surface and covered with polyethylene. Application should be made on a mockup panel or small samples prior to use on finished architectural panels.

When chips or spalls are to be repaired, all loose material should be removed using a hand-held or pneumatic chipping hammer. The areas should be cut back, usually 1 to 2 in., (25 to 50 mm) until coarse aggregate breaks under chipping rather than being loosened. The sides of the hole should be essentially vertical or slightly undercut to provide a key at the edge of the repair. Depth of area to be repaired should be a minimum of 1.5 times the maximum aggregate size. Very thin edges (feather edges) should not be permitted; cutouts should be at least 3/4 in. (20 mm) deep to provide a shoulder of sufficient depth to permit a smooth finish. For exposed aggregate finishes, the cutting-back of the concrete in an irregular pattern minimizes detection of the repaired area. In liner-type finishes it may be desirable to cut back along the lines of the texture to obscure the location of the repair.

Pre-mixed amounts of dry repair materials should be formulated at the plant to be used in the field as required. Structural products may be repaired with a commercially pre-mixed cement-based patching material or with a job-mixed sand-latex mortar mix. Commercial products are available in varying degrees of setting speed with the faster setting speeds used to "build up" large patches and the slower setting speeds used for finishing.

After the hole is cut and dressed, all traces of loose debris and dust should be removed by brushing or by blowing with compressed air. Then the area to be repaired and an area at least 6 in. (150 mm) wide surrounding it should be thoroughly saturated with water to prevent absorption of water from the repair mix into the surrounding concrete and to facilitate good bond with repair materials. If a bonding agent is used, the area is left dry. After surface water has evaporated (approaching dryness) from the area to be repaired, a 1/16- to 1/8-in. (1.6 to 3 mm) bond coat of neat cement slurry, mixed to the consistency of thick cream, or a bonding agent should be thoroughly brushed into the surface. When the bond coat begins to lose the water sheen, the pre-mixed repair mixture should be applied.

Bonding agents such as epoxies, polyesters and water-resistant latexes (ASTM C1059, Type II)[40] may be desirable for large repairs. Latex bonding agents should generally not be used integrally with the repair material, since a different texture and color as well as a different appearance (when wet) will be apparent. These differences will not fade with time due to different absorption characteristics (or the repair may even change to an objectionable color). The bonding agent should be neatly applied to the dry surface of the repair area and allowed to dry to a tacky-sticky condition before applying the repair mix. Care should be taken to prevent the bonding agent from overlapping the edge. It may be desirable to hold back bonding agent 3/4 in. (20 mm) from the finish surface. Very dry or very absorptive concrete may require a second application of bonding agent. If, inadvertently, an epoxy bond coat reaches a soft rubber-like stage (no tackiness) prior to the placement of the repair, a second application of the epoxy bond coat should be applied.

Rubberized or plastic buckets that are easy to clean should be used for mixing. Each batch should be dry-mixed to thoroughly blend the materials. Care should be taken to avoid spilling any of the mixture during blending because this can change proportions and cause color variations. Water should be added slowly while hand-mixing. For best control of consistency, just enough water should be added so that the mix will hold together upon rolling into a ball, making sure there are no dry materials left at the bottom of the container. Unmixed materials may cause premature stiffening and inadequate color blend. If water is added too quickly, the mortar may be too wet and have to be discarded.

As a final step in the mixing process, the trowel/slump test should be used to measure batch consistency. A quantity of mortar is lifted on the back side of the trowel and the trowel tapped on the side of the mixing bucket twice. When the trowel blade is turned to vertical, the mortar should stick to the trowel. If the mortar slips off, it is too wet and the batch should be discarded. If the mortar falls off the trowel, it is too dry. The water should be adjusted as needed to achieve exactly the correct consistency.

After completion of mixing, the mixing container should be covered with a damp cloth to minimize surface evaporation and the mix should be undisturbed for 5 minutes to allow for moisture absorption. Also, when not being applied, the mix should be kept covered with a damp cloth to minimize surface evaporation and be frequently manipulated with a trowel, without addition of water. This procedure, combined with proper curing, will minimize the formation of drying shrinkage cracks. Repair mix batches should be used within one hour.

The repair mix should be stiff but not so stiff that aggregate cannot be pressed into it. A small margin or pointed plastic trowel or dense sponge float should be used to apply and compact the repair mix into final position. Do not use a metal trowel because this can darken the surface, especially on white or light-colored concrete. The dark color is nearly impossible to remove. Also, wire brushes or wire instruments of any kind should not be used because they can result in metal scrappings or particles embedded in the repair which may later appear as rust spots.

Repairs should be made out of direct sunlight. Methods such as sunshades should be considered to protect repairs from the heat of the sun and to keep the repaired area cool before and after repairs.

The repair mix should be placed in equal layers not more than 1/2 in. (13 mm) thick. Each layer should be given a scratch finish to improve bond with the subsequent layer. Each layer should be placed immediately after the consolidation of the previous layer. The repair mix should be either vibrated, if possible, or vigorously hand-tamped for consolidation and to assure positive and uniform contact with the previous layer. The repair layers should be tamped with 1 to 2 in. (25 to 50 mm) round or rectangular plastic or wood rods. Wooden tamping rods should be sealed with three layers of a urethane form sealer. This will minimize possible discoloration, absorption or sticking of the repair mix to the tamping rod. Vibration or tamping will bring water to the surface which will assist in finishing the relatively dry mixture.

The mix should be applied and packed to the approximate density of the surrounding concrete. Variation in density will result in different absorption characteristics. If left unsealed, repairs that are of a different density than surrounding areas will be noticeable during periods of wet weather or if a staining sealer is used.

When the repair is deeper than 2 in. (50 mm) or the repair extends more than 2 in. (50 mm) in from a panel edge, supplemental reinforcing may be required. This also applies to large spalls which are greater than 2 in. (50 mm) in depth. Holes should be drilled for stainless steel dowels at 9- to 12-in. (230 to 305 mm) centers along the length of the repair. Dowels should be anchored with epoxy. Additional reinforcing steel or welded wire reinforcement may be added and tied to the dowels, as necessary.

The surface of the repair should be struck-off using a wooden or plastic trowel, leaving the repair slightly higher than the surrounding surface. To permit initial shrinkage, the repair should be left undisturbed for at least 1 hr. or until initial set of the repair mix occurs. After initial set, the repair surface should be restruck flush with the adjacent concrete and smoothed with the trowel. The finished surface should be level and flush to minimize shadow effects. Excess mortar should be removed with a damp, soft sponge or with a stiff, nylon bristle brush and water to obtain the desired surface texture.

As an alternative to the placing procedure described above, drypacking may be used where appropriate. Drypacking, i.e., the placement of a very stiff mortar by ramming or tapping into place, should be used for deep holes which are not excessive in area, for rock pockets, and for any hole which has a relatively high ratio of depth-to-area. Drypack should not be used for shallow depressions where lateral restraint cannot be obtained, for filling-in behind long lengths of exposed reinforcement, or for filling holes which extend entirely through concrete sections. If these conditions are satisfied, the drypack method of repair will usually be satisfactory.

Most drypack is a mix of 1 part cement to 2 1/2 parts sand (by dry volume) that will pass a No. 16 (1.18 mm) sieve. The amount of water to be used should be only enough to produce a mortar which will stick together on being molded into a ball by a slight pressure of the hands and will not exude water but will leave the hands damp. The proper amount of mixing water and the proper consistency will produce a filling which is at the point of behaving rubbery when the material is packed solidly. Any less water will not make a sound, solid pack; any more water will result in excessive shrinkage and a loose repair. The cavity should be cleaned of loose material and kept damp with water for several hours. A bonding grout (1 part cement, 1 part fine sand passing a No. 30 (600 µm) sieve, and enough water to form a creamy consistency) should be scrubbed with a brush onto all surfaces to which the mortar is to be bonded. The grout should not be allowed to dry before the mortar is placed. Drypack mortar should be placed and packed in layers having a compacted thickness of about 3/8 in.(10 mm). Each layer should be solidly compacted over its entire surface by use of a hardwood or plastic stick and a hammer. Metal tampers are not suitable, as they tend to polish the surface, thus making the bond to successive layers less satisfactory.

After initial set, the surfaces of repairs should be dressed manually to obtain the same texture as surrounding surfaces. The methods applicable for achieving the proper texture are dependent on the type of finish.

A repair on smooth finish is very difficult to disguise because of variations in color and texture, particularly if dark cements are used. Color differences in dark concrete are more noticeable than in lighter-colored concrete. A smooth surface may be obtained by wiping the repair with cheese cloth or a similar soft material, rubbing with carborundum stone, or by grinding. A light abrasive blast with very fine silica blasting sand may remove the dark color. This abrasive blast is applied using a suction fed sandblasting pot with a trigger-type nozzle at 40 psi (276 kPa) maximum pressure. Sandblast approximately 28 days after applying the repair. Also, if the surface has numerous bugholes, it may be desirable to simulate the bugholes with the point of a nail prior to initial set of the repair concrete.

In those cases where a liner texture was applied, it may be necessary to use a liner to impress onto the repair the texture desired as the concrete sets. The compacted repair should be struck flush with the original surface. A small section of formliner which has been attached to a board or wooden float should be pressed against the repair while the repair mix is plastic. If possible, the liner should remain in place to help with curing. Where the surface is board-marked, the texture can be carried across a repair by striking-off the surface with a straight-edge held parallel to the direction of the marks. If the surface is ribbed, it may be necessary to form the rib and fill the form with the repair mix. The length of the ribs should be varied so that a straight line pattern does not extend across the ribs.

For exposed aggregate finishes, after the compacted repairs are struck flush with the existing surface, damp coarse aggregate should be hand-placed into the mix, if not already added in the repair mix itself. The coarse aggregate should be patted into the surface with a small wooden block and a wood or rubber float should be used to cover the aggregate with the repair mix. After initial set of the repair mix, stiff nylon bristle brushes and water may be used to expose the aggregate and texture the finish to blend with the surrounding concrete. A wet paint brush or soft natural sponge may also be useful to achieve the desired texture. Sponging the repair surface after texturing can be effective in removing cement laitance and may be adequate to eliminate the need for further cleaning of the repair. Care should be taken to not roll the aggregate and damage the fresh repair mix material.

To obtain a sandblasted finish, the repair should be allowed to cure for 5 to 7 days, then lightly sandblasted using a suction-fed sandblast pot with a trigger-type nozzle at 30 to 60 psi pressure, depending on the depth of exposure. A fine-graded silica blasting sand, 120 to 200

(125 to 75 μm) mesh size, should be used. A conventional sandblaster with a 3/8 to 3/4 in. (10 to 19 mm) diameter venturi nozzle may also be used, provided that a pressure valve in the air line maintains the pressure at the desired range. Another method to obtain a deep sandblasted finish is to allow the repair to cure for 7 days, then use a deep blast on the repair, filling-in with the proper matrix and then lightly brushing to proper depth prior to initial set. In some cases, the aggregate may be sandblasted prior to use in the mix. This will abrade and dull the surfaces of the aggregate to match the surfaces of the exposed aggregate in the original, surrounding surface.

If honed or polished surfaces must be repaired, the repair should be built-up 1/8- to 3/16 in. (3 to 5 mm) higher than the surrounding finished face, then ground flush using hand tools of successively finer stone until the specified finish, level and planeness have been achieved. Small chips should be repaired using a color-tinted epoxy containing crushed and graded coarse aggregate fines to minimize the paste volume. Repairs should be hand-polished 24 hours after application to match the surrounding concrete.

Adequate curing methods for repairs should be implemented immediately to ensure that the repair does not dry too quickly and cause it to shrink away from the existing concrete. The repair area should be moist-cured for a minimum of 3 days with 7 days more preferable. Moist curing may be accomplished by covering with a blanket of damp flannel, linen, cheese cloth or other non-staining fabric and a cover of polyethylene taped or tied to the concrete. Care must be used to avoid washing or eroding the repair. If needed during the curing period, additional water should be added to the fabric and the plastic covering replaced. In any event, the curing method should be tested on the trial repairs made in the mockup sample unit. Non-staining membrane curing compounds may be sprayed over a repair and later removed by scrubbing with soap and water or by acid washing, if the unit was so treated. Heat lamps that will maintain a temperature in the repair of between 50 to 75 deg. F (10 to 24 deg. C) are recommended for curing in cold weather.

After all repairs have been completed, they should be coated with a silane or dilute (2 to 3 percent solids) solution of an acrylic sealer to minimize moisture migration into the repair concrete. The acrylic sealer should not stain the concrete. Even with proper consolidation, the repair is more porous than the original hardened concrete and will need this diluted application of sealer.

Cleaning should not be started until 3 to 7 days after repairs. Regardless of the method used, extreme caution is necessary to ensure that the repair is not overworked or discolored. In some cases, cleaning of repair areas with a brush or with a brush and concrete cleaner may be sufficient to remove objectionable laitance. Acid cleaning may be used if the rest of the unit was so treated.

Corrosion-protected materials should be touched-up upon completion of all intended curing and acid cleaning.

If minor damage occurs to veneer stone during shipping, handling or erection, successful remedial field work can be accomplished. Such repairs are normally done by the precaster with repair procedures developed in consultation with the stone supplier. Epoxy, stone dust and a coloring agent, if necessary, can be used to repair small chips or spalls. These repairs can be finished to the same surface texture as the stone facing. If it is necessary to replace a section of stone, satisfactory techniques have been developed in both cases when the back of the panel is accessible and after the panel has been erected and the back of the panel is inaccessible.

If minor damage occurs to a clay product facing during shipping, handling or erection, remedial field repairs can be accomplished easily, including replacement of the individual clay products. However, because clay products of one color will not match all of the various hues in use on any one job, it is important to determine the quantity of repairing that needs to be done to result in an acceptable appearance.

When required, fins or burrs left during manufacture of structural members may have to be removed and patched to a smooth surface. Burr removal can be done most efficiently with a right-angle hand grinder fitted with a cup grinding wheel or a #50 grit sanding disc. The same grinder fitted to a wheel mounted "giraffe" can be used to dress ceilings if a significant number of burrs must be removed. The grinder mounting to the giraffe neck should be a controllable articulating connection, permitting the operator to maneuver the grinder head to any position desired.

The repaired area should conform to the balance of the work with respect to appearance, strength and durability. Ideally, repairs should not be evaluated for acceptability until they have cured for 28 days. Repaired samples should be viewed at a distance or height consistent with actual panel locations. Also, a panel should be viewed in its entirety from a distance of 20 ft. (6.1 m), and not examined for individual deficiencies. The concrete surface should not be wet at the time of evaluation.

Cracks

If cracks have occurred, and if repairs are required for the restoration of structural integrity or aesthetics, the cracks that range in width from 0.002 in. to 0.25 in. (0.05 to 6 mm), may be repaired by pressure injection of a low-viscosity, high-modulus, 100-percent solid, two-component epoxy conforming to ASTM C881.[41] Type, grade, and class should be chosen to satisfy job conditions and requirements. The system should be capable of bonding to wet surfaces unless it can be assured that the crack is completely dry. Care should be taken to select an epoxy color (amber, white or gray) which most closely matches the concrete surface.

Small cracks, equal to or less than 0.010 in. (0.26 mm) in width, may not need repair, unless failure to do so can cause corrosion of reinforcement in severe environments. If repair is required for the restoration of structural integrity, cracks may be pressure injected with a low-viscosity epoxy that will bond to a moist surface because it is usually impossible to remove moisture in the crack. The general procedures involved in epoxy injection are as follows (see Refs. 42 and 43).

1. *Clean the cracks.* Proper preparation of the crack area is most important. Form release agents, efflorescence, grease, dirt, or fine particles of concrete prevent epoxy penetration and bonding. Preferably, contamination should be removed by vacuuming or by flushing with water or other effective solvent. The solvent is then blown-out using compressed air or adequate time is provided for evaporation.

   Crack areas in a finished face to receive a sandblast texture should not be sandblasted prior to epoxy injection, since this will round the edges and accentuate the crack.

2. *Install entry ports.* Entry port holes for the injecting process should be drilled into and along the length of the crack. The size of the drill bit selected is determined by the size of the crack, both width and depth, and by the injection port selected. Through cracks in any precast concrete unit should have injection nipples installed in both crack faces at alternating locations. This means the spacing on each side will be 2 times the member thickness. For members 1 ft (0.3 m) or less in thickness, ports should be placed in the crack on one side only and spaced at the thickness of the member. However, if the cracks are less than 0.005 in. (0.13 mm) wide, entry ports should be spaced not more than 6 in. (150 mm) apart. This will ensure that epoxy injected into one port will flow through the full depth of the crack before flowing out through the next port. The first and last entry ports should be established at or near the bottom and top, respectively, of any vertical crack, or at the ends of any horizontal crack in a vertical or horizontal member.

   Three methods of installing the entry ports are in general use:

   a. *Drilled holes with fittings inserted.* This method is often used in conjunction with the grooving of cracks. When grooving, the top edges of the cracks are cut to a V-shape approximately 1/4 in. (6 mm) deep. A hole is drilled into the crack, approximately 3/4 in. (19 mm) in diameter and 1/2 to 1 in. (13 to 25 mm) deep below the apex of the V-grooved section. A fitting, such as a pipe nipple or tire valve stem, is bonded into the hole with an epoxy adhesive. A vacuum chuck and bit are helpful in preventing the cracks from being plugged with drilling dust.

   b. *Bonded flush fitting.* When the cracks are not V-grooved, a method frequently used to provide an entry port is to bond a fitting flush with the concrete surface over the crack. This flush fitting has a flanged cylinder cross section with an opening at the top for the adhesive to enter.

   c. *Interruption in seal.* Another system of providing entry is to omit the seal over a portion of the crack. This method can be used when special gasket devices are available that cover the unsealed portion of the crack and allow injection of the adhesive directly into the crack without leaks.

3. *Seal the surfaces.* Once the ports are cemented in place with a surface sealant and before injection, the entire crack surface should be sealed by brushing an epoxy, polyester, or other appropriate sealing material along the surface of the crack both front and back (or top and bottom) and allowing it to harden to keep the epoxy from leaking out before it has gelled. If unexposed faces of the concrete cannot be reached, crack repair by pressure injection is extremely difficult unless special steps are taken. If extremely high injection pressures are needed, the crack should be cut out to a depth of 1/2 in. (13 mm) and width of about 3/4 in. (19 mm) in a V-shape, filled with an epoxy, and struck off flush with the surface.

   Low-pressure injection permits the use of easily-removed silicone sealants, hydraulic cement, polyester, hot wax, a strippable thermoplastic seal, or a variety of other single-component materials to cap the crack. A silicone can be peeled-away and the other low pressure seal applications removed with modest abrasion, heat, or chemicals, leaving an architectural surface without significant blemishes. These sealants tend to be fragile and require special control of injection pressures. The surface sealant should cure before beginning injection of the epoxy.

4. *Mix the epoxy.* The epoxy may be mixed either by batch or continuously. A color toner may be used in the epoxy to match adjacent concrete surfaces. In batch mixing, the adhesive components are premixed according to the manufacturer's instructions, usually with the use of a mechanical stirrer, similar to a paint mixing paddle. Mixing of epoxy may be accomplished by hand for quantities less than 1 qt (0.001 m^3). Care should be taken to mix only the amount of adhesive that can be used prior to gelling. When the adhesive material begins to gel, its flow characteris-

tics begin to change, and pressure injection becomes more and more difficult. In the continuous mixing system premixing is not necessary as the two liquid adhesive components pass through metering and driving pumps prior to passing through an automatic mixing head. The continuous mixing system allows the use of fast-setting adhesives that have a short working life.

Similar to the reaction between cement and water, heat is given off when the A (epoxy resin) and B (curing agent) components of an epoxy system react. In a thin film on a cold substrate, the heat generated will not be noticeable.

All two-component epoxy systems are temperature-sensitive. The colder the environment, the slower the material cures, with different formulations or product lines having various lower limits to the temperatures at which the reaction ceases. Few epoxy resins cure satisfactorily at ambient or concrete temperatures below 50 deg. F (10 deg. C). High temperatures [generally those above 85 deg. F (29 deg. C)] should be avoided because increased temperature increases the rate of hardening and makes it difficult to inject the epoxy resin material before it gels. Some formulations cure at reasonably low temperatures and, if frozen, won't expand and spall concrete sections but lie dormant until the temperature rises enough to allow the reaction to resume.

Prior to mixing, the directions on the labels should be read carefully. Ensure that the shelf life of the product has not expired, and that both the A and B components are in the correct proportions and are meant to be blended together (i.e., both components are identified by the same product trade name or number). Mix in containers that the material is packaged in, or otherwise in clean, dry containers. If required, sand should be added after the A and B components have been thoroughly blended. Pure epoxies (no sand added and no solvent in formulation) have a pot life of from 12 to 20 min. at 70 deg. F (21 deg. C). If large quantities of sand are added, the pot life may be doubled. If large amounts of solvent are in the formulation, pot life can range from 30 min. to 8 hrs. The pot life can be extended by cooling the material in a cold water or ice bath, but the viscosity will change and application may become difficult. Conversely, a faster cure (shortened pot life) will result if the epoxies and sands are heated before being applied to the repair areas. All tools and equipment should be kept continuously clean with solvents.

5. *Inject the epoxy.* Hydraulic pumps, paint pressure pots, or air- or hand-actuated caulking guns can be used. The pressure used for injection should be carefully selected. The pressure is usually in the range of 10 to 90 psi (69 to 621 kPa). Increased pressure often does little to accelerate the rate of injection. In fact, the use of excessive pressure can propagate the existing cracks, causing additional damage.

Low pressure systems generally use prebatched resin rather than depending upon a dispenser for metering. Immediately following mixing, the resins are pressurized by springs within capsules, inflatable syringe-like devices, or air. In each case, reservoirs of limited capacity are used. Therefore, the resin supply should be carefully monitored and replenished as necessary or an incomplete glue line may result.

There is frequently a temptation to terminate low-pressure injection prematurely once the initial observable flow has ebbed. This is a mistake for two reasons. First, the resin flow is not always noticeable since the flow rate may be as low as one or two ounces per hour on small cracks. Yet two ounces of resin can fill a great deal of crack and be absolutely essential to the finished repair. Secondly, though the crack may be filled, pressure should be maintained to compensate for absorption, especially in the case of dry and porous concrete. This differs from high-pressure injection where the concrete within the fault is normally super-saturated by the injection pressure, eliminating any likelihood of subsequent absorption. The best practice is to leave low-pressure systems in place until the resin has gelled.

The use of long pot life (LPL) resins is absolutely essential for successful injection with low pressure systems. A gel time of one hour at 72 deg. F (22 deg. C) is recommended.

If the crack is vertical or inclined, the injection process should begin by pumping epoxy into the entry port at the lowest elevation until the epoxy level reaches the next highest entry port above. Starting at the lowest port helps to vent trapped air and moisture through the higher ports. The lower injection port is then capped, and the process is repeated at successively higher ports until the crack has been completely filled and all ports have been capped. For horizontal cracks, the injection should proceed from one end of the crack to the other in the same manner.

If the epoxy begins to leak from a hole in the surface sealer, the injection process should stop. The leaky surface should be heated with a propane torch or hot air heating gun until the resin just starts to bubble. The heating should then cease and the epoxy allowed to cure for approximately two minutes before restarting injection. The crack is not considered full if the pumping pressure cannot be maintained because the epoxy may be flowing into unfilled portions of the crack or leaking out of the crack.

During the injection operation it is very difficult to be sure that the crack is completely full. The experience of the applicator is very important. Although some work has been done on sophisticated methods, such as ultrasonic testing to determine whether or not a crack has been filled, the only practical method available is by drilling concrete cores. This may be necessary when complete assurance of a sound structural bond is required.

The use of epoxy injection is not applicable for sandwich panels in areas where insulation is located because once the epoxy penetrates the injected wythe, there then exists an unlimited path of travel for the epoxy at the concrete to insulation interface. "Painting" of the cracks with injection epoxy is sometimes used to seal the cracks.

6. *Remove the surface seal.* After 3 to 24 hours, the temporary surface seal and ports are scraped off leaving the face of the concrete smooth and unmarred except for the trace outline of the crack. The epoxy should be examined for degree of hardening. It may be necessary to texturize the crack area with a needle-gun to match the surrounding concrete surface. A follow-up light sandblast at low pressure (less than 50 psi) can be used in final clean-up.

7. *Clean-up.* Before the epoxy becomes tack-free, all tools should be wiped clean with paper towels and immersed or cleaned with mineral spirits, toluene, methyl ethyl ketone (M.E.K.) or xylene. Each epoxy compound contains some materials that are toxic before cure. Skin contact with uncured material may cause mild to severe dermatitis with some people if their skin is exposed to the resins for long periods of time. Also, precautions should be used when working with powerful solvents such as M.E.K. or xylene. Long-sleeved clothing, rubber gloves and safety goggles should be worn. If contact occurs, wash immediately with soap and water. Avoid inhalation of vapor and use proper ventilation. Clean rubber gloves or disposable polyethylene gloves should be used for hand protection. Contact with the eyes must be avoided but if contact should occur, flush immediately with water for 15 minutes and call a physician. *Caution: All safety labels and Material Safety Data Sheets should be read before using solvents or epoxy resins.*

If a loose, spalled piece of concrete is retrieved and surfaces mate, the easiest repair is to simply reattach the piece using nonsag adhesives or bonding agents. Broken surfaces of the unit as well as the loose piece should be painted with the adhesive. Enough epoxy should be applied to the surfaces to ensure that some will squeeze out of the joint when mated pieces are clamped together. The epoxy consistency should be selected to prevent sagging or running on a vertical surface. Self-leveling formulations should only be used on horizontal surfaces. In some cases it may be necessary to secure the repair by inserting pins in epoxy-filled holes drilled in the loose pieces and into the precast concrete unit.

APPENDIX F — JOINT SEALANTS

The importance of well-designed and installed jointing systems cannot be overemphasized. Joint design and treatment is critical to the performance of the finished structure. Joints are required to accommodate changes in precast unit or structure dimensions caused by changes in temperature, moisture content or loads. Joint sealants (caulking) must at the same time prevent water and air penetration through the building envelope. The joint treatment also has an effect on the general appearance of the project. For a comprehensive discussion of joint sealants, refer to ASTM C1193.[44]

Joint Design

Sealant life and performance are greatly influenced by joint design. For optimum performance and maximum life, the recommendations of the joint sealant manufacturer should be followed. Joints between precast units should be wide enough to accommodate anticipated movements. Particular attention should be given to joint tolerance (see Article 5.3) in order for the joint sealant system to perform within its design capabilities. If units cannot be adjusted to allow for proper joint size, enlargement by saw cutting may be necessary.

When joints are too narrow, bond or tensile failure of the joint sealant may occur or adjacent units may come in contact and be subjected to unanticipated loading, distortion, cracking or local crushing (spalling).

The required sealant depth depends upon the sealant width at the time of application. The optimum sealant width/depth relationships should be based on recommendations of the sealant manufacturer. However, because all manufacturers do not agree on proportions, generally accepted guidelines are:

1. For joints up to 1/2 in. (13 mm) wide: The sealant should have a minimum depth of 1/4 in. (6 mm) and a maximum depth equal to its width.
2. For joints from 1/2 in. to 1 in. (13 to 25 mm) wide: The sealant depth should be equal to one half the width, but not less than 1/2 in. (13 mm). The sealant should have a concave shape providing greater thickness at the panel faces.
3. For joints in excess of 1 in. (25 mm) wide: Sealant depth should be limited to 1/2 in. (13 mm) maximum. For sealant width exceeding 2 in. (50 mm), the depth should be determined by consultation with the sealant manufacturer.

Joint Preparation and Sealant Application

The sealants used for different specific purposes are often installed by separate subcontractors. For example, the window subcontractor normally installs sealants around windows, whereas a different subcontractor typically installs sealants between panels. The designer should select and coordinate all of the sealants used on a project for chemical compatibility and adhesion to each other. In general, contact between different sealant types should be minimized.

The recommendations of the sealant manufacturer should always be followed regarding surface preparation, priming, mixing, application procedure and application life. Good workmanship by qualified sealant applicators is the most important factor required for satisfactory performance.

The edges of precast concrete units should be sound, smooth, clean and dry. They should be free of frost, dust, loose mortar or other contaminants such as form release agents, retarders, or sealers that may affect adhesion. It may be more economical and effective to prepare joint surfaces prior to erection if a large number of units need surface preparation.

Sealants which cure chemically should not be applied to wet or icy surfaces as they may cure or set before they can bond to the concrete surface. Some methyl methacrylate resin sealers inadvertently sprayed in the joints may peel away from the concrete surface leaving a void between joint sealant and concrete, while silicone water repellents in the joints will prevent adhesion of joint sealant to concrete surface. Use of a stiff wire brush, light grinding or sandblasting followed by air blowing may be necessary to remove surface contaminants.

Also, before caulking, the joint should be wiped with a cloth dampened with an oil-free solvent such as xylol and given a dry wipe. Sometimes, smooth concretes which are very shiny exhibit a "skin" on the surface. The skin may peel leaving a gap between it and the concrete after the joint sealant has been applied to the concrete. It may be necessary to remove the skin with a stiff wire brush followed by a high-pressure water rinse.

To enhance adhesion between sealant and concrete surface, a primer may be recommended by the sealant manufacturer. Special care should be used to avoid staining the visible face of the precast unit because some primers leave an amber colored stain if brushed on the surface. The primer should be allowed to cure before application of the sealant. The sealant and primer should always be supplied by the same manufacturer. In some circumstances, it may be desirable to conduct adhesion or peel tests to determine the compatibility of the sealant with contact surfaces prior to proceeding with the job.

The depth of the sealant should be controlled by accurate placement of a suitable backup material. To obtain the full benefit of a well-designed shape factor, the backup materials should have a bond breaker surface or be flexible enough that it doesn't limit shape change of the sealant. However, many backup materials do not adhere to sealants and thus, where these are used, no separate bond breaker is needed. In addition, a bond breaker strip should be used in all joints where there is insufficient room to place backup material, to ensure free movement of the sealant.

When selecting a backup material or bond breaker, the recommendations of the sealant manufacturer should be followed to ensure compatibility. The backup should not stain the sealant. Sometimes, it may bleed through and cause discoloration of the joint. Backup materials should be of suitable size and shape so that, after installation, they are compressed 30 to 50 percent. Proper selection and use of backup material is essential for the satisfactory performance of watertight joints. Length-wise stretching, twisting or braiding of the tube or rod stock used for backup should be avoided. When inserting a polyethylene foam backup rod, a blunt tool should be used to avoid skin puncture of the rod and possible out-gassing which may cause blistering of the sealant.

The caulking gun should have a nozzle of proper size and should provide sufficient pressure to completely fill the joint. Joint-filling should be done carefully and completely by thoroughly working the compound into the joint. Upon being completely filled, joints should be tooled and left slightly concave to eliminate air pockets or voids, and to provide a smooth, neat appearing finish. The surface of the sealant should be a full, smooth bead, free of ridges, wrinkles, sags, air pockets and embedded impurities.

A tooling agent may be used to help reduce drag in the tooling process. Water is probably the best agent, however some applicators prefer soapy water. If soapy water is used, it should be used with caution as it can contaminate joints not yet sealed and affect adhesion adversely.

When tooling light-colored sealants, it is recommended that tools be used dry, or if wet, water should be clean. Other solvents may be suitable as tooling agents, provided they are specifically approved by the sealant manufacturer.

Surfaces soiled with sealant materials should be cleaned as work progresses, as removal is likely to be difficult after the sealant has cured. A solvent or cleaning agent that is recommended by the sealant manufacturer should be used.

Considering moisture condensation at low temperatures and reduced working life at high temperatures, a practical range of installation temperatures is from 40 to 90 deg. F (4.4 to 32 deg. C). However, it is recognized that applications may have to made at lower temperatures. When this is necessary, steps must be taken to ensure clean, dry, frost-free surfaces. The area to be sealed should be wiped with a quick-drying solvent such as xylol, just before sealing. Also a wider than specified joint may have to be formed, or changes in the type of sealant or depth-to-width ratio may be required to secure greater extensibility. The applicator should know the joint size limitation of the sealant selected. The area may be heated, if possible, or at least the sealant should be slightly warm when applied.

Sealing of joints in exposed aggregate panels requires the sealant to be applied against relatively smooth concrete because good adhesion is difficult to achieve against an exposed aggregate surface. The sealant should be held back from the face of the exposed aggregate and that portion of the matrix along the joint should present a smooth, clean surface for the application of the caulking material. This requirement is simple to comply with for recessed joints. Also, for maximum performance, sealants should not be applied to beveled surfaces, but should be applied beyond the beveled area.

When a two-stage joint is used, proper venting and draining is required. Venting prevents water (either from penetration or condensation) from collecting between the wall structure and facing panels or within the panel joints, if either or both the exterior and interior faces of the panels are sealed. Vent tubes should be 1/4 to 3/8 in. (6 to 13 mm) inside diameter, polyvinyl chloride or other non-staining material. The vent tube should slope down to the exterior face of the panel and must penetrate the joint backing to allow free movement of air between the outside environment and the cavity. A minimum of two vents per panel, spaced approximately 6 to 10 ft (1.8 to 3 m) on centers should be used. All horizontal joints should have vent tubes, and in most cases these vents would be located at the junction of the horizontal and vertical joints. Vents should project at least 1/8 in. (3 mm) beyond the sealant.

When installing a two-stage joint, the interior air seal should be applied first. This will prevent warm, moist air from the interior entering the joint cavity, which could cause condensation.

Certain joint designs and inaccessible joints after erection may require the use of preformed gaskets. The gasket material may be pre-glued to the precast unit to fix its position. Air seal material should be applied reasonably straight (especially if exposed on the interior of the panel), but should not be stretched during gluing or installation operations. Air seal gaskets should be spliced at 45 deg. in ship-lap fashion. The gasket material should protrude beyond the precast unit sufficiently to ensure a reasonably tight butt joint following installation of the precast members.

Neoprene compression seals should remain compressed approximately 15 percent of the uncompressed seal width at the maximum joint opening to maintain sufficient contact pressure to seal and resist displacement. Generally, the amount of compression is limited to 55 percent of the uncompressed seal width at maximum closing to prevent overcompression. For larger sizes of compression seals, the 55 percent value may be exceeded. The allowable movement is thus approximately 40 percent of the uncompressed seal width.

The allowable movement for impregnated foams is much less than for neoprene, about 5 to 20 percent.

It is recommended that pre-glued gaskets, not be used in vertical joints because proper compression of the gasket during erection may be difficult to achieve. However, if required, it is usually necessary to move the precast unit horizontally with come-a-longs after release from crane and initial fastening of top connections.

In order to facilitate installation, the recommendations of the manufacturer of the compression seals should be followed relative to lubricants, adhesives and conditioning of materials in cold weather.

When installing a fire-resistive joint sealing system, the fire resistive blanket should be installed with a minimum 10 to 15 percent compression.

Expansion joints (isolation joints) are provided in parking structures to reduce the stresses associated with volume changes. The most common method of sealing these joints is a factory-molded urethane sealant installed over a traffic plate by adhering into place with polymeric nosings. Deterioration of underlying structural elements and leakage can be expected when these expansion joints are not properly sealed.

Structures must often be built with control joints and construction joints. Topping pours must have control joints over joints between the precast units. These joints should be tooled into the plastic concrete to assure proper joint action. Sawn-joints should not be used because cracking may occur before sawing takes place and because a tooled V-joint provides a better sealant seat. A high-quality traffic-bearing polyurethane sealant should be used to prevent intrusion of salts into these joints and to prevent subsequent deterioration of embedded metals. As a backup in case of sealant failure, all metal passing through the joint should be epoxy-coated, galvanized or coated with a zinc-rich paint.

When the underside of hollow-core slabs is to be left exposed to view, the bottom of the grout key may be caulked. The horizontal gap between the hollow-core slab and the wall at the bearing ends may also be caulked. Sealant should not be applied to the horizontal gap between the slabs and the wall along the non-bearing sides as the up-and-down-movement of the slabs at this location will frequently rupture the sealant bead.

APPENDIX G — CLEANING

When testing appropriate procedures for the removal of dirt, stains and efflorescence, a suggested order is (beginning with the least damaging):

1. Dry scrubbing with a stiff fiber nylon brush, particularly if the surface is brushed shortly after the appearance of efflorescence.

2. Abrasive blasting with industrial baking soda to remove efflorescence. This abrasive material will not affect the concrete surface. Any residue on the surface should not be removed by water as salts will be dissolved and carried into concrete causing additional efflorescence. Residues should be blown, vacuumed or brushed from the surface.

3. Wetting the surface with water and vigorous scrubbing of the finish with a stiff fiber brush followed by thorough rinsing of the surface with clean water. Low-pressure water spraying (water misting), high-pressure water or steam cleaning may also be tried.

4. Chemical cleaning compounds such as detergents, muriatic or phosphoric acid or other chemical cleaners used in accordance with the manufacturer's recommendation. If possible, a technical representative of the product manufacturer should be present for the initial test application to ensure its proper use.

   Areas to be cleaned chemically should be thoroughly saturated with clean water prior to application of the cleaning material to prevent the chemicals from being absorbed deeply into the surface of the concrete. Surfaces should also be thoroughly rinsed with clean water after application so that no traces of chemicals remain in the surface layers of the concrete. The cleaning solution should not be allowed to dry on the concrete finish. Residual salts can flake or spall the surface or leave difficult stains. Misapplication of hydrochloric acid can lead to corrosion of embedded metals that have shallow concrete cover.

   Care should be taken to use dilute solutions of acid to prevent surface etching that may reveal the aggregate and slightly change surface color and texture. The entire unit should be treated to avoid a mottled effect. Any of several diluted solutions of acids are effective ways to remove efflorescence:

   a. one-part hydrochloric (muriatic) acid in nine- to 19-parts water
   b. one-part phosphoric acid in nine-parts water
   c. one-part phosphoric acid plus one-part acetic acid (vinegar) in 19-parts water
   d. one-part acetic acid in five-parts water

   Hydrochloric (muriatic) acid may leave a yellow stain on white concrete. Therefore phosphoric or acetic acid should be used to clean white concrete.

   Workers using cleaning compounds or acid solutions should be thoroughly trained in their use. The use of proper protective wear should be strictly enforced.

   Rubber gloves, glasses and other protective clothing must be worn by workmen using acid solutions or strong detergents. Materials used for chemical cleaning can be highly corrosive and frequently toxic. All precautions on labels should be observed because these cleaning agents can affect eyes, skin and lungs. Materials which can produce noxious or flammable fumes should not be used in confined spaces unless adequate ventilation can be provided.

5. Dry or wet abrasive blasting, using sand or other abrasives, may be considered if this method was originally used in exposing the surface of the unit. The architectural precaster or an experienced subcontractor should be engaged for sandblasting. A venturi-type nozzle should be used on the gun for its solid blast pattern, rather than a straight bore nozzle that produces light fringe areas.

6. Stone veneer-faced precast concrete units should be cleaned with stiff fiber, stainless steel or bronze wire brush, a mild soap powder or detergent and clean water using high pressure, if necessary. Acid or other strong chemicals that might damage or stain the veneer should not be used. Information should be obtained from stone suppliers on methods of removing oil, rust and dirt stains from the stone.

7. Mortar stains may be removed from brick panels by thoroughly wetting the panel and scrubbing with a stiff fiber brush and a masonry cleaning solution. A prepared cleaning compound is recommended; however, on red brick, a weak solution of muriatic acid and water (not to exceed a 10 percent muriatic acid solution) may be used. Acid should be flushed from the panel with large amounts of clean water within 5 to 10 minutes of application. Buff, gray or brown brick should be cleaned in accordance with the brick manufacturer's recommendations, possibly using proprietary cleaners rather than acid to prevent green or yellow vanadium stains and brown manganese stains.

   Following the application of the cleaning solution, the panel should be rinsed thoroughly with clean water. High-pressure water cleaning techniques, using a 1,000 to 2,000 psi (6.9 to 13.8 MPa) washer, may also be used to remove mortar stains.

   Unglazed tile or terra cotta surfaces should be cleaned with a 5-percent solution of sulfamic acid for gray or white joints, and a more dilute (2 percent) solution for colored joints. The surface should be thoroughly rinsed with clean water both before and after cleaning. Glazed tile manufacturers generally do not recommend the use of acid for cleaning purposes.

8. For information on removing specific stains from concrete, reference should be made to "Removing Stains and Cleaning Concrete Surfaces," IS 214, published by the Portland Cement Association, Skokie, IL.[45]

APPENDIX H — SURFACE SEALERS AND CLEAR COATINGS

Coatings for Horizontal Deck Surfaces. Concrete surface sealers reduce moisture and salt (chloride) penetration, which is particularly important in parking structures. While these sealers can enhance the durability of any concrete topping, they do not substitute for basic durable concrete design, nor do they provide protection against penetration of moisture and chlorides through cracks. Research has shown that the performance of concrete sealers varies, depending on the product and other variables. Products should be evaluated against the criteria established in the NCHRP 224 study.[46]

Sealers may be classified into two groups: penetrants and surface sealers.

Penetrants. These are generally silanes or siloxanes. They penetrate the surface, reacting with cementitious materials, making the concrete hydrophobic. They do not have crack-bridging capabilities. These materials do not appreciably affect the appearance or characteristics of the surface to which they are applied. While generally more expensive than other types of sealers, they are typically longer lasting and less subject to wear under traffic or deterioration from exposure to sun. For optimum wear, forty-percent solids usually is recommended.

Surface Sealers. These are generally polymer resins such as urethanes, epoxies, acrylics or other proprietary blends. They protect by penetrating the surface slightly, and/or by providing a tough continuous film over the surface; these materials also do not bridge cracks. Surface sealers are generally less expensive than penetrants and performance characteristics of many of these products compare favorably with the penetrating sealers, as demonstrated by the NCHRP 224 criteria. These sealers are more likely to be subject to wear under traffic and may be more slippery than the bare concrete surface.

Clear Surface Sealers for Vertical Surfaces. Clear surface coatings or sealers are sometimes used on precast concrete units to improve weathering qualities or to reduce attack of the concrete surface by airborne pollutants. Because of the quality of the concrete normally achieved in plant-cast precast concrete, even with very thin sections, sealers are not required for waterproofing. Because the results are uncertain, use of sealers in locations having little or no air pollution is not recommended.

A careful evaluation should be made before deciding about sealers. This includes consultation with the local precasters. In the absence of near-identical experience, sealers should be tested on reasonably sized samples of varying age to verify performance over a suitable period of exposure or usage based on prior experience under similar exposure conditions.

Sealers should be applied in accordance with the manufacturer's written recommendations. Any clear sealer used must be guaranteed by the supplier or applicator not to stain, soil, darken or discolor the finish, cause joint sealants to stain the precast concrete surface or affect the bond of the sealant. The manufacturers of both the sealant and the sealer should be consulted before application, or the materials specified should be pretested before application.

The effectiveness of sealers is dependent upon the properties of the specific sealers. Some problems that have occurred with certain sealers are:

1. Appearance varies with the age of the precast unit being treated. If a methyl methacrylate resin sealer of high solids content is applied, the member will take on a "wet" look. Some sealers may create a blotchy appearance if applied before the units are fully cured and dried, while others may lose their effectiveness if applied too soon after fabrication. Sealers may also accentuate a patched area of different density. If it is necessary to remove sealers, use a solvent or wire brush, grind or lightly sandblast the surfaces.

2. Certain sealers, such as silicones, have been found to attract airborne hydrocarbons.

3. Sealers will often interfere with patching of the precast concrete surface or with adhesion of joint sealants. Some methyl methacrylate resin sealers inadvertently sprayed in the joints may peel away from the member surface leaving a void between sealant and unit while silicone water repellents in the joints will prevent adhesion of joint sealant to the concrete surface. This is why the sealer/sealant compatibility should be verified.

4. The possibility of severe and permanent discoloration of the panel surface to various shades of yellow,

brown, or gray, or of peeling, varies considerably between types and sources of sealers. Moisture permeability (breathing) of the sealer is a requirement to prevent blistering and peeling of the sealers.

5. Proper application, following the sealer manufacturer's instructions, depends on qualified operators and possible expensive pretreatment of the precast concrete units. The application limitations of sealers with respect to timing, ambient temperatures, moisture content of the concrete, method, and rate of application should be fully investigated before choosing a particular type of sealer or supplier.

6. Uncertain life expectancy, possibly causing a maintenance expense through resealing.

The type of solvent used in sealers, as well as the solids content, can affect the resulting color of the concrete surface. Thus, neither the type nor the source of sealer should be changed during the project. Generally, sealers having higher solid contents tend to darken the concrete surface. The amount of color change depends primarily on the type of material in the sealer and the concentration, as well as the porosity of the surface. The active ingredient in the sealer should be chemical-resistant to the alkaline environment of the concrete. Also, sealers should be evaluated on how well they penetrate precast concrete surfaces that vary in absorption and texture. The penetrating sealers, generally silanes or siloxanes, develop their water repelling ability by penetrating the surface to depths of 1/4 in. (6 mm), reacting with the cementitious materials in the concrete and rendering the concrete hydrophobic. The penetrating ability of the sealer system depends on the molecular size of the active ingredient, the viscosity of the system, and the solvent-carrying system.

In most cases, except for the silane or siloxane sealers, it may be necessary to adequately roughen an as-cast surface in order to obtain good adhesion of the coating. One way is to lightly sandblast the surface to be coated.

Low-pressure, airless spray equipment should be used to apply sealer. This results in a uniform application while avoiding excessive rundown. Two coats are usually required to provide a uniform coating because the first coat is absorbed into the concrete. The second coat does not penetrate as much and provides a more uniform surface color. Care should be taken to keep sealers off glass or metal surfaces.

Jobsite applied surface coatings should not be applied until all repairs and cleaning have been completed. In cases where the precast concrete units have been coated at the manufacturing plant, and additional cleaning is required, it may be necessary to recoat those particular members.